Expert Systems
in Developing Countries

Expert Systems in Developing Countries

Practice and Promise

EDITED BY

Charles K. Mann
and Stephen R. Ruth

Routledge
Taylor & Francis Group

LONDON AND NEW YORK

First published 1992 by Westview Press

Published 2018 by Routledge
52 Vanderbilt Avenue, New York, NY 10017
2 Park Square, Milton Park, Abingdon, Oxon OX14 4RN

Routledge is an imprint of the Taylor & Francis Group, an informa business

Copyright © 1992 by Taylor & Francis

Library of Congress Cataloging-in-Publication Data
Expert systems in developing countries : practice and promise / edited
   by Charles K. Mann and Stephen R. Ruth.
      p. cm.—(Westview special studies in social, political, and
economic development)
   Includes bibliographical references and index.
   ISBN 0-8133-8397-8
   1. Expert systems (Computer science)—Developing countries.
I. Mann, Charles Kellogg, 1934–   . II. Ruth, Stephen R.
III. Series.
QA76.76.E95E9768   1992
630′.2′085633—dc20   91-39502
                                                                   CIP

ISBN 13: 978-0-367-01663-0 (hbk)

ISBN 13: 978-0-367-16650-2 (pbk)

Contents

v

Acknowledgments

As with any work of relatively broad scope, our debts in bringing this book to completion are many. First, we thank the American Association for the Advancement of Science for the opportunity at their 1988 annual meeting to launch the project with a panel discussion: The Use of Expert Systems in Developing Countries. Panel member James W. Jones was especially helpful in assisting us in tracking down additional examples of expert systems in Third World agriculture. Chapter authors greatly facilitated our editorial task by their thoughtful and prompt responses to suggestions, aimed at making the overall work more coherent and consistent. In particular, Victor S. Doherty contributed not only as chapter author but as a refreshing source of insights and stimulus. Jerry Mechling at Harvard's J. F. Kennedy School of Government and John J. Sviokla at the Harvard Business School also provided valued encouragement and insight.

For administrative assistance and painstaking care in reworking notes and references into a consistent format, we are grateful to Amy Ganders, Patricia Carney, and especially to Jennifer Gonzalez who assisted in the critical final stage of manuscript preparation. For bringing the manuscript through many drafts into camera-ready copy, we express special appreciation to Mary Blackwell, Director of George Mason University's Word Processing Center, and to Sandy Slater, who not only did a superb job with the text processing but also converted many of the graphics directly into camera-ready form.

At Westview Press, we thank Kellie Masterson for her early encouragement to us to develop this book. As the manuscript progressed, Lynn Arts and Ellen McCarthy provided excellent guidance to assure a consistent format and appearance for all of the chapters. For the comprehensive index, we are indebted to Veronica Cruz. Lastly, we thank our respective institutions, the Harvard Institute for International Development and George Mason University School of Business Administration, for providing an environment to encourage reflection as to how the cutting-edge technologies of the industrialized countries may be shaped to benefit the developing countries as well. Finally, we close this statement of appreciation and acknowledgment with our acceptance of final responsibility for any errors or omissions that may yet remain.

Charles K. Mann
Stephen R. Ruth

1

Introduction and Overview

Charles K. Mann and Stephen R. Ruth

Two Visions of the Future

"The expert systems whose stories we have told are windows on the new economic results from the first technology spin-off of artificial intelligence. Even in the light of early dawn, the returns on investment, the productivity increase, the consequences of preserving and distributing company expertise, the enhancement of quality, and the other gains are remarkable. In the international competitive environment, the early adopters will not easily be displaced by latecomers. The competitive opportunity presents itself now. When will another, as widely useful and powerful as this, come along?" (Fiegenbaum, McCorduck, and Ni 1988:270,71)

"...There is almost no likelihood that scientists can develop machines capable of making intelligent decisions. After 25 years of research, AI has failed to live up to its promise, and there is no evidence that it ever will. In fact, machine intelligence will probably never replace human intelligence simply because we ourselves are not 'thinking machines'. Human beings have an intuitive intelligence that 'reasoning' machines simply cannot match.... Computers that 'teach' and systems that render 'expert' business decisions would eventually produce a generation of students and managers who have no faith in their own intuition and expertise." (Dreyfus and Dreyfus 1986:44)

Expert Systems in the Developing Countries

An expert system (ES) is a computer program that uses knowledge, facts, and reasoning techniques to solve problems that normally require the abilities of human experts. The goals sought by expert system builders include helping human experts, assimilating the knowledge and experience of several human experts, training new experts, and providing requisite expertise to projects that cannot afford scarce expertise on site.

The two statements that open this chapter illustrate the wide spectrum of opinion that exists about the future of expert systems. One finds views ranging from unbounded optimism to pessimism that few practical benefits soon will be realized. If the future now appears so uncertain in the developed countries, why then expend energy and intellect on expert systems in the developing countries? There are at least six reasons.

First, the pessimism about the current use of expert systems in the developed countries turns out to be pessimism only in contrast to the exaggerated expectations aroused in the decade of the eighties. There are many useful, successful applications of expert systems in use, some of which are described briefly in the chapters of this book.

Secondly, the wide variety of inexpensive, off-the-shelf expert system development packages brings the technology to a wide universe of potential ES developers. Much as the microcomputer empowered a whole new set of software developers, expert system "shells" dramatically expand the talent pool of people able to work creatively on a vast array of potential applications. In effect, the expert increasingly is able to become his or her own knowledge engineer. The numbers of people working with the technology should accelerate the use of the technology.

Thirdly, given that the ES technology now runs on the existing large and rapidly growing microcomputer base that exists in the developing countries, ES applications—unlike the microcomputer itself—need not originate only in the developed countries. Using the MS DOS computers now common in almost all developing country technically oriented organizations, plus an expert system shell costing less than the spreadsheet or wordprocessing programs already installed, the Third World may be as likely a locus of ES applications as the First World.

Fourth, since many of the everyday problems and solutions of the Third World have not assumed the same degree of complexity as in the First World, there may be a substantial number of applications where relatively small, simple expert systems can make an extremely useful contribution. For example, the village health diagnostic system developed for an Ethiopian refugee camp's conditions can address 80% of the problems presented by providing diagnostics and treatment regimes for only 5 disease complexes (Chapter 12).

Fifth, in the developed countries with their extensive vocational training and engineering training capacity, there are large numbers of well-trained, mid-level technicians. There is a large amount of expertise relatively widely embodied in the existing stock of human capital. Moreover, with good transportation and communication systems, experts are more easily accessed. In the developing countries, in contrast, there is a relatively thin layer of expertise at the top, a huge unskilled labor force at the bottom, and a very thin layer of technicians in the middle. Poor transport and communications systems makes the expertise that exists more difficult to access. The potential of expert systems to capture relatively simple levels of expertise may be of far greater use in the developing countries than in the developed ones. To some extent they may be able to substitute for the missing middle level of technician, allowing less well-trained persons to perform technical tasks assisted by an ES.

Lastly, in the developed countries with their good transportation and communications systems, experts not only are more plentiful, they are easier to reach. Consider, for example, computer technical assistance "hotlines" and telephone call-in service from extension agents, to name only two examples. In the developing countries access to such experts as exist is more restricted by poor transportation and communication systems.

For all of the above reasons, it seems likely that expert system applications may play some important roles in the developing countries relatively early in their evolution. Moreover, in many cases, the tasks to which they are applied may be quite different from applications in the developed countries. Indeed, the considerations listed above suggest that they should be quite different.

Despite the fact that expert systems in general are being taken up more slowly than their original enthusiasts expected, the reasoning above suggests that the development community should not wait until ES is a relatively mature technology in the developed countries before exploring possible applications in developing countries. Accepting this argument, the American Association for the Advancement of Science agreed to sponsor a panel in 1988, organized by Ruth and Mann, on applications of expert systems in developing countries. The presentation and discussions at that panel served as the starting point for the development of this book. Assisted by our fellow panelists, especially Victor Doherty, we sought out the experience of others working in this subject to add additional perspectives to this book and to make more concrete some of the points made above.

Partly to provide greater coherence to the book, but mainly because many of the pioneering applications of ES in development relate to agriculture, eight of the chapters focus on ES in agricultural development. On the demand side, agriculture dominates economic activity in many of the developing countries and also is a major focus of agencies providing development funding. On the supply side, agricultural engineering faculties are an important component of US land grant universities and are a major force in developing ES applications. Many of these institutions are importantly involved in development abroad and are quick to spot potential ES applications there. Lastly, various aspects of agricultural expertise fit well the profile of expertise suitable for embodiment in expert systems. (See Jones, p. 34, 35)

Many of the people who attended the AAAS panel came not because of their extensive knowledge about expert systems, but because they worked in development and wanted an introduction to expert systems in the context of developing country applications. Assuming that some of our readers also fit this profile, there follows a brief introduction to ES. Following this introduction, brief chapter overviews present the overall plan of the book and highlight several recurring themes.

The Development of Expert Systems

Expert systems developed in the 1960s and 1970s were typically written on a mainframe computer in the programming language based on List Processing (LISP). Evolving from university research laboratories, they were limited to the applications developed by these research sites. Most of these expert systems were not intended for commercial use. They incorporated the specific knowledge of the experts about the problem area, termed "domain knowledge," problem-solving heuristics (or "rules of thumb") and inferencing capabilities, and an interface mechanism between the user and the system. Some examples of these systems include MACSYMA, developed at the Massachusetts Institute of Technology (MIT), for assisting individuals in solving complex mathematical problems; Stanford University's MYCIN, which diagnosed bacteremia and meningitis infections, and the University of Pittsburgh's INTERNIST/CADUCEUS, which aided internal medicine diagnosis and decision making.

Later researchers at Stanford University realized that MYCIN consisted of three distinct parts which can be visualized as set of concentric circles or as a seed. At the kernel of the seed is a knowledge base which contains the domain specific knowledge. Outside that is the inference engine, the part which contains the inferencing capabilities, problem-solving heuristics, and control strategies. And finally, the expert system is "surrounded" by the end user-system interface. By removing the domain-specific knowledge, and by adding tools for managing knowledge sets (such as rule editors and tracers), these scientists created a general-purpose tool for developing expert systems, now called a "shell."

Expert System Shells

It was not until the 1980's that commercial shells were introduced for a variety of classes of computers. Some of the more popular shells today are 1st-Class, ADS, ART, CRYSTAL, EXSYS, GoldWorks, Guru, Level 5,

Nexpert Object, KDS, KES, M.1, Personal Consultant, S.1, TIMM, and VP Expert. These packages not only provide the necessary tools for developing an expert system such as the user-system interface, inference mechanism, rule editor, and code optimizer, but can generally be run on microcomputers in addition to minicomputers and mainframes, are reasonably priced, and provide powerful features without requiring an individual to learn the mechanics of an Artificial Intelligence (AI) language or purchase specialized hardware. With the development and widespread use of shells, the range of expert systems applications has tremendously increased. Their introduction has played a major role in expanding expert system applications into areas such as management, finance, office automation, computer selection and networking, legal processes, manufacturing, equipment training, personnel training, education, transportation, oil and geology, science and medicine, and agriculture.

Overview of the Book

Following this introductory chapter, in Chapter 2, Steve Ruth compares the costs and benefits of high-end expert systems with shell-based systems. The rapidly increasing power of these shell systems running on personal computers makes them an increasingly attractive alternative to high-end systems. It also makes them an appropriate means by which an organization can explore potential advantages of an expert system without making a major commitment to hardware and software dedicated only to that application. In addition, shell-based systems often have the added advantage of being able to draw the organization's existing staff resources directly into the creating of expert systems. This helps to build capacity within the organization rather than relying upon outside expertise. However, just because shell based systems are relatively low in cost does not mean that they will prove cost effective in all applications. Ruth provides a list of issues to consider before investing in expert system development.

Moving from these general considerations to the context of agricultural development, Victor Doherty in Chapter 3 identifies a range of roles for expert systems. Like Professor James Jones in Chapter 4, he sees ES contributing importantly to transfer of expertise from the agricultural scientist and research station to the extension staffs. However, unlike Jones who does not see a large role for ES at the farm level in developing countries, Doherty argues that the technology is rapidly becoming sufficiently ubiquitous that it may play an important role on relatively small farms in the developing countries. Lastly, Doherty presents the provocative and imaginative idea that an expert system could be developed to embody the decision rules and "expertise" of the developing country farmer—the target audience of the research and extension service. This would allow the researcher to have immediately available at all stages in technology development an ES proxy for "the farmer"; perhaps even a socio-economic range of farmers. Continued interaction with the ES "farmer" would help the researcher to see the world as the farmer sees it; would help him to filter research ideas through the filters the farmer is using.

In Chapter 4, James Jones describes the characteristics of agricultural knowledge, arguing that agriculture is particularly well-suited to the application of expert systems. He sets the stage for chapters to follow by categorizing existing ES applications into four broad groups: the production rule approach; production rules linked with simulation model; frame based systems; decision support systems combining a several interacting approaches with one or more expert system modules. To make the categories more concrete and understandable, he provides examples of existing agricultural applications within each group. With the rapid spread of PCs he sees ES playing an increasingly important role in the representation, storage and transfer of knowledge.

To introduce the reader to some of the key concepts underlying most expert systems, Thomas Fermanian and Ryszard Michalski describe in Chapter 5 AGASSISTANT, an expert system builder for agricultural applications. This system is of particular interest as it is an extension and expansion of Michalski's PLANT/ds, one of the first expert systems in agriculture and the first agricultural expert system that learned its rules from examples. Running on a VAX minicomputer, PLANT/ds was developed to help farmers and extension agents correctly diagnose soybean diseases commonly found in Illinois and continues to be used for this purpose. Now expanded and generalized to handle a wide range of agricultural problems, AGASSISTANT runs on the IBM PC and compatibles.

AGASSISTANT's inferencing mechanism was specifically designed to combine levels of uncertainty commonly found in agricultural domains. It also provides a set of tools for the development of a knowledge base through learning by example. Of special interest is the discussion of the inference engine at the heart of the advisory system and description of the four different ways the system can develop the rules upon which the inference engine operates. In effect, the system is capable of learning its decision rules from example, and of improving its knowledge as it "acquires experience." To demonstrate how AGASSISTANT can be applied to a specific problem, the chapter concludes with the presentation of WEEDER, an expert system developed to help identify the species of grass weeds present.

In Chapter 6, Pierce Jones and his colleagues address one of the key challenges in developing an expert system: the effective extraction of knowledge from experts. The authors sum up the problem with a quote from AI pioneer Waterman: "The more competent domain experts become, the less able they are to describe the knowledge they use to solve problems." Using a case history of the development of an expert system for making spray recommendations for pests in soybeans, they offer practical advice and insights on working productively with subject matter experts to make explicit the essence of the expertise. The chapter is also useful in demonstrating how an experienced extension specialist's expertise was incorporated into an expert system. In a developing country setting, where extension specialists are far scarcer than they are in the US, the ability to codify and replicate the expertise of these few specialists would be extremely useful. Another line of thought stimulated by this chapter relates to Victor Doherty's intriguing idea of trying to embody a typical village farmer's "expertise" within an expert system for the benefit of researchers. In this respect, the chapter provides a checklist of procedures to help elicit from the farmer information on how he or she perceives and approaches various problems.

Illustrating a system designed explicitly for extension agents, in Chapter 7, Rafea, Warkentin, and Ruth describe the creation of an ES in Egypt. The subject domain is the production of cucumber seedlings in protected plastic tunnels. Particularly useful to those considering an expert system is the discussion of criteria to determine the technical feasibility of a proposed ES application. In terms of Jones's categories of systems, this application illustrates the use of a system combining both rules and frames. The authors also explore issues of prototype design and development and implementation issues.

In Chapter 8, Professor Warkentin and colleagues describe UNU-AES, the United Nations University Agroforestry Expert System. This system is intended to assist individuals in applying agroforestry approaches to land management for sustainable production of food and fuelwood supplies by farmers in developing countries. It focuses on alley cropping, a type of agroforestry system in which leguminous trees are planted in rows with food crops cultivated between them. The system design team set as its performance criteria for the system that it would give the same advice as a recognized expert on alley cropping. The chapter describes how questions to users are formulated, the system advice options that depend upon answers to these questions, provides a representative decision rule within one of these options, and illustrates a sample of the sort of advice given to the user by AES. At this stage in its development, the system does not include data, knowledge or rules relating to socio-economic variables that might condition recommendations. However, the authors note that such considerations can and should be built into subsequent versions of the system. It could also be expanded to include other agroforestry technologies in addition to alley cropping.

In contrast to the UNU-AES, Chapter 9 illustrates a system that does attempt to incorporate socio-economic considerations into the recommendations it generates for various categories of farmers. The context is Indonesia, where many persons have been encouraged to migrate from highly populated Java to the less populated outer islands. The acid upland soils of these islands require very different soil management techniques from those appropriate to the rich volcanic soils of Java. Russell Yost and his colleagues at the University of Hawaii describe their work in developing expert systems that can provide advice appropriate not only to particular soil types, but advice appropriate to the situations of farmers from differing socio-cultural backgrounds. The design team was highly interdisciplinary and included an anthropologist whose task was to work on these issues of socio-economic "fit." The case study is particularly useful in demonstrating both the practicality of incorporating farmer knowledge and beliefs within the system, as urged by Doherty in Chapter 3, but also in illustrating some of the formidable challenges encountered in doing so.

Chapter 10 demonstrates how an expert system can help to integrate the the spatial dimension into the planning of agricultural development. David Mendez and Severin Grabski describe DREAGIS, a hybrid combination of an expert system with a geographic information system, developed to help improve land use in the Dominican Republic. Representing an example of Jones's fourth category of systems, DREAGIS illustrates how an expert system can enhance the benefits derived from other complementary computer information systems. In addition to information on how the system was conceptualized and implemented, the authors offer some provocative insights on how the system has forced recognition of the lack of correspondence between the extension service regions and agro-ecological regions. The construction of the system also facilitates asking "what if" questions fundamental to analyzing a range of policy options.

As did Warkentin and colleagues in Chapter 8, Mendez and Grabski note the importance of adding economic considerations to the system to move beyond an analysis of what is feasible to an anlysis of what is profitable, and ultimately, what combination of crops is optimal, according to some specified objective function. Bringing in the socio-economic characteristics of the farmers themselves is not yet included in the future work plan. Perhaps some of the lessons learned in the Indonesian soil management system described in the previous chapter will generate some ideas as to how to approach this dimension of the task. By the same token, the GIS work described here would seem to have important applications to the soil managment system.

Chapter 11 shifts the subject matter domain from agricultural development to health care in the developing countries. John Daly presents a comprehensive framework within which to consider the generation of information technology and its transfer and use within developing countries. While his focus is on applications in primary health care, the framework is robust and general enough for application to a wide range of fields.

In considering prospects for the diffusion of information technology, he examines systematically the factors likely to affect its spread. He categorizes these as scientific, technological, institutional, economic, social, and political determinants. Supporting his reasoning with evidence from the historical development and diffusion of medical technology, he foresees substantial under-investment in health-related information technology in developing countries, particularly as it could benefit the poor. "Physicians and hospitals probably will be more successful in appropriating microcomputer and micro-electronics technology than paraprofessionals and primary health services, and therefore will exacerbate a trend toward high cost hospital-based medical services. The potential in computer technology to de-skill health care probably will be resisted effectively by physicians, due to their political and organizational authority."

To reduce the likelihood of the negative outcomes he foresees, he urges improvement within the developing countries themselves of the capacity to develop, adapt, introduce and analyze the impact of computer technologies in primary health care. This includes involving in health technology decisions nationals who embody the relevant moral, ethical, and religious values; seeking ways to subsidize research on computer technology for preventive and primary health care; donor support to projects emphasizing computers to support and help in training paraprofessionals and other lower and mid-level manpower including mid-wives, Ayurvedic practitioners, and pharmacy clerks. Echoing Doherty's inclusion of the farmer himself or herself as a potential end-user of the technology, Daly also urges subsidies for development and application of computer technology for the use of families and individuals in developing countries in meeting their own primary care needs.

Chapter 12 demonstrates in concrete terms how a research subsidy can support the development of an expert system to help train and support village health workers, in this case a system tailored for conditions in a part of Tigre Province in northern Ethiopia. On the one hand, this application demonstrates how powerfully such a system can improve the quality and quantity of health care provided by relatively unskilled village personnel. On the other hand, it demonstrates how area-specific and circumstance-specific such a system must be to work effectively. It must reflect the local disease distribution (in this case five core disease areas account for most of the health problems), geographic area (highland), political context (civil war), social environment (poverty, famine), infrastructure of health care (rural health centers closed, two town hospitals), capacity of users. The authors conclude that "It is not possible to devise a detailed generic general purpose knowledge-based system suited for village health workers in different countries. For its proper function and acceptance, the system has to be tailored both to the region where it is used and the people using it." Moreover, even this relatively well-developed system, the authors do not consider yet ready for field testing, due largely to its lack of a user interface appropriate to the educational level of the village health workers.

Clearly there is an enormous way to go before expert systems can move into widespread use to improve primary health care. Nonetheless the existence of this system demonstrates clearly that—given the sort of support that Daly calls for—systems can be developed to improve substantially village level health care. At the same time, the extent of tailoring to local conditions suggests the magnitude of the investment needed, at the current state-of-the-art, to develop systems to empower a significant share of the world's village health workers with this technology.

In contrast to the expert system aimed at improving village level health care, Eckhard Kleinau explores in Chapter 13 how an expert system might be used to help decision makers weigh options for the financing of health care at the national level. Because of growing financial constraints and increased demands on national health systems, developing country governments increasingly are moving from free or near free health services to some sort of cost sharing with the patient. However, in addition to direct payments by the patient for care, there is a wide array of other financing mechanisms. Kleinau argues that policy makers do not consider all of the possible financing options in part because of the complexity of doing so. The expert system that he proposes would help policy makers deal with this complexity. It would help them to see both the financial implications of alternative cost recovery schemes and also to judge the likely effect on demand for services and upon the health status of the various groups in the population.

The ES that Kleinau proposes is intended to complement rather than replace existing computer models. Generally these deal with discrete parts of the problem; they explore one particular financing mechanism. Thus, part of the attraction of an expert system is its potential in helping to integrate information from multiple sources, and thus to allow the user to compare various combinations of cost recovery schemes. Another appealing feature is the interactive nature of a good ES and its ability to "explain" its reasoning to the user. These features allow the ES to become an important capacity building tool, helping the user to understand better how various types of cost recovery affect the financing of health care and the mechanisms by which each option affects various target populations.

In Chapter 14 Metka Vrtacnik and colleagues describe an expert system used to expedite responses to water pollution emergencies on major rivers in Czechoslovakia. The system helps to identify pollutants, to determine effective responses, to identify the source of the pollution, and to provide advice on reducing the pollution at its source. The system features use of two relational databases with commercial expert system shells. A particularly interesting feature of this system is the provision for rapid and frequent updating through subscription to a commercial database of chemical substances.

The examples in Chapters 3–14 notwithstanding, expert systems applications are still relatively rare in the developing countries. However, in Chapter 15, Charles Mann argues that an expert system "perspective" can produce substantial benefits by encouraging people to think about existing knowledge systems in new ways. In the case of the *AskARIES Knowledgebase*, applying an expert system perspective to an existing database produced large benefits, even though no expert system was involved. The stimulus came from a shift in the organizing metaphor. The metaphor underlying a database is the relatively passive one of a filing system. This does not tend to stimulate a wide range of questions about the contents of the files. Shifting perspective to the metaphor underlying an expert system—the activist, problem-solving expert—fundamentally alters the way one thinks about the database and its contents. Instead of a focus on the filing system within which the information is organized, the focus shifts to more active analysis of the content of the files and to a more interactive relationship between system and user.

Approaching a database as the potential knowledgebase component of an expert system can generate fresh insights as to possible "rules" or "lessons" that might be derived from the data. Also, such a shift in the organizing metaphor can suggest more effective ways to organize the information (e.g. around clusters of problems) and can help in developing a more accessible and interactive user interface.

In developing the *AskARIES Knowledgebase*, shifting perspective from database to knowledgebase also suggested a whole new information collection strategy. Whereas the original conception relied only on "the literature" of small-scale enterprise development, the new focus suggested adding information from in-depth interviews with experts in the subject, including their "rules of thumb" for dealing with various problems. The mindset of "annotated bibliographic database" did not suggest talking to subject domain experts. Shifting to the mindset of "knowledgebase" within a nascent expert system opened a whole new avenue along which to seek solutions to the problems of developing small enterprise programs.

The book ends with reflections on future promise in light of current practice. To illustrate, Victor Doherty, Charles Mann, and John Sviokla note how expert systems, once they are in place, could serve to speed adoption of new agricultural technology. By providing a framework into which new expertise easily could be incorporated, they promise to make ever more rapid the adoption of technology. This final chapter also suggests how continued miniaturization, combined with cellular and digital communications, may expand the usefulness of expert systems by allowing hand-held field systems to be linked to such real time information sources as national weather and market information services. The chapter closes with the observation that expert systems not only can serve to deliver expertise to a wide range of users, but also can make important contributions to capacity building through their ability to "explain" how they reach their conclusions.

References

Dreyfus, H. and S. Dreyfus. January, 1986. Why computers may never think like people. *Technology Review*.

Feigenbaum, E., P. McCorduck, and H. Penny Ni. 1988. *The Rise of the Expert Company: How Visionary Companies are Using Artificial Intelligence to Achieve Higher Productivity and Profits*. New York: Times Books.

2

Harnessing Shell-Based Knowledge Engineering Environments: An Applications-Based Perspective

Stephen R. Ruth

Introduction

One of the key issues limiting the establishment of knowledge based software programs in large organizations is the apparently prohibitive cost of the investment. It is often necessary to question whether the return on investment for such systems can be obtained through reduced operating cost, enhanced services, higher long run quality of performance or other expected favorable results in traditional profit centers. And the typical expenses associated with any new introduction of technology, fixed costs as well as variable costs, invariably come under scrutiny in the context of yet another information technology. This chapter briefly examines the way costs of knowledge-based systems are calculated and segregated and then introduces a low cost, high yield approach to the process of developing knowledge-based systems that can result in dramatic reductions in costs.

Costs of Knowledge-Based Systems

This cost has a variety of manifestations but can be conveniently divided into four major components: software expense, hardware expense and the expenses of knowledge elicitation resources and appropriate domain experts. Each of these components can account for major costs lodged against the project budget. It is not unusual for such budgets to be in the millions of dollars annually (Bobrow, Mittal, and Stefik 1986; Sviokla 1990).

---

*Segments of this article were presented at U.S. Postal Service Advanced Technology Conference, November 5, 1990.

The promise of expert systems, so widely heralded in the literature of medicine (Kunz 1978), agriculture (Lemmon 1986), business and accounting (Engming 1986), education (Ruth 1990), and many others, comes at a price that can be daunting for many organizations. Some of the most famous ES implementations, like MYCIN and XCON, have been characterized by large expenditures—and large payoffs. Particularly in the case of XCON, the Digital Equipment Corporation invests several million dollars per year—but gains bottom line savings of tens of millions annually (Sviokla 1990)

Software Cost

The reason that such large—and successful—systems have been very costly is closely related to the complexity of the delivered software. The components of this software are frequently described as these three elements (Prereau 1990, 17):

1. A knowledge base, for storing the essential facts, vocabulary, and rules;
2. An inference engine for causing these facts and rules to be linked to the expert's method of applying decision rules. (the term "firing" is used to refer to a decision by the inference engine to use a particular rule or rule sequence)
3. A dialog system to connect users as well as system developers to the system.

Each of these components is a complex computer program written in powerful languages like LISP, PROLOG or in ES development environments like R1 or KEE. It is not unusual for such programs to consist of thousands of individual instructions each. Using a generally accepted heuristic that each such instruction costs about a day's pay for a programmer (Mahler 1986), it is easy to accumulate life cycle costs in the range of several million dollars for a large program written in this way. For example, assuming that each component required about three thousand instructions (a rather conservative estimate), the rough life cycle cost would be in the range of:

3 components x 3,000 instructions per component x $300 daily rate for journeyman programmer = $2.7 million

Even if the daily rate were far less, there is little doubt that software expense can easily be a million dollar charge against any potential gains from the project.

Hardware Cost

The powerful, so-called high end expert systems described above require sophisticated computers for maximum performance. Many of these systems need minicomputers called LISP Machines or could even require mainframe computers in some cases. To establish and maintain an environment for the required computers requires annual expenditure of several hundred thousands of dollars.

For example, one LISP machine with its support staff allocation and required peripheral devices can easily reach $100,000 per year, after the initial investment, which is in the realm of $70,000. If a mainframe computer is selected, it would typically be running the expert system for only a small percentage of its billable time. Yet the costs allocated for this more limited use of a much larger resource would in aggregate be at least as large as for the LISP machine.

Costs of Domain Experts and Knowledge Engineering Specialists

There are relatively few knowledge engineers currently available compared to the demand for their services. Depending on the type of organization these specialists may have titles that range from futuristic to traditional: knowledge elicitor, systems designer, programmer, systems engineer, systems analyst and knowledge engineer (Mc Graw and Harbison-Briggs, 1989, 5). By whatever title, the special skills required in this area are costly. Similarly, the domain experts, the persons who are to be the source of the knowledge-based systems, are typically individuals whose specialties are well compensated. Annual costs for two knowledge engineers full time and two domain experts part time can easily exceed several hundred thousand dollars.

An Alternative Approach: Low Cost, High Yield Expert System Shells for Focused Problems

An alternative to the powerful but costly methodology just described is the use of an emerging type of software that combines the three elements of the expert system into a developmental environment requiring a small fraction of the programming expense and yet offering many of the advantages of the larger, more complex environments. These ES "shells" are like tool kits for developing expert systems and come with pre-written software for inference engine, knowledge base and dialog system. A recent article (Ruth and Carlson 1989) described these tools across several dimensions. Table 1 gives an updated and revised summary of the differences between the large scale tools and ES shells.

Shells vs. High End Systems: A Comparison

As the table indicates there are vast differences between the high end systems and shell-based systems. Several of these differences deserve more careful attention. The fact that shell-based systems run on personal computers and are about one tenth of the unit cost of the high end environments could be attributed to a drastic difference in methods of utilization of the two environments. Popular shells like EXSYS, LEVEL 5 Object and 1STCLASS provide a full range of graphics, the ability to chain to other programs (including

FIGURE 2.1. Brief Comparison of Selected Variables Associated with High End and Low End ES Implementation Environments

Characteristic	High End	Low End
Typical Life Cycle Cost	$1–10 Million	$10,000–100,000
System Development Elapsed Time	Years	Months
Number of Rules or Rule Equivalents	5,000–50,000	200–5,000
Typical Computer Environment	Minicomputer; Mainframe	Personal Computer
Typical Unit Cost	$200–500/rule or rule equivalent	$20–100/rule or rule equivalent
Examples of Software Used	LISP, PROLOG R1, KEE	Shells (EXSYS, 1STCLASS, LEVEL V, others)
Acquisition Cost of Software	$50,000–250,000	$100–5,000
Locus of Program Development	Information Systems Group	IS Group or User Group
Reported ROI of the Most Successful Systems	10–1	10–1
Estimated number of Systems Attempted	Hundreds to Thousands	Hundreds of Thousands

spreadsheet and database software), can be fully integrated into minicomputer or mainframe processing and can incorporate demonstrations or methods as necessary. Shells are obviously easier to use, since they include the inference engine, dialog system and knowledge base software already programmed and the user/programmer need only insert the logic and dialog messages as necessary.

With respect to the number of rules, shells are well below the capacity of high end systems, but it should be observed that a large number of high end systems result in several thousand rules, well within the capability of the more robust shells. Hence, the advantage of large rule capacity is gained only in systems of very high complexity—and correspondingly high expense.

The most compelling statistic is the life-cycle cost of shells compared to high end systems. While there is ample evidence that unit costs are significantly lower for shells the degree of that difference is not clear. A rough heuristic used by Dr. Ed Mahler at duPont is that shell based systems require an hour's work per rule while high end systems need about a day's work per rule (Kupfer 1987; Mahler 1986). This closely matches Gavarter's estimate of approximately an order of magnitude advantage in unit costs in favor of shell based systems. (Gavarter 1987)

Determination of Leverage Points

Because shell-based ES development is low in unit cost there is a possibility that an organization may consider developing ES applications without considering the feasibility, size of gain, strategic value and other issues that should be part of any information system venture (Leonard-Barton and Sviokla 1988) They may simply plunge in, since the down-side risks are relatively low. This procedure is usually not successful (Sheil 1987). Instead, each organization should examine the potential leverage points offered by the system, just as would be the case in more in the case of more complex environments. Some significant questions are:

- Will the spreading of expertise in this very limited area enable us to reduce contingency funding?
- Are we in danger of losing this expertise?
- Do we need a more consistent approach throughout the organization in this decision area?
- Is the information codified somewhere else but hard to access?
- Will this ES affect our effectiveness in accomplishing our mission?
- Will this application broaden the skills and potential of our employees?

Agenda for Action

If an organization is contemplating an investment in some knowledge based activity the comparisons offered in this paper should suggest several insights that could be of help in planning for most efficient use of scarce financial resources.

Phased Approach

Taking advantage of knowledge based technology demands a careful, incremental approach to utilization. It would seem that shell based approaches would be an appropriate method of initial entry. Once a series of systems are developed using lower cost shell tools a decision can be made as to the appropriateness of more advanced tools. This approach offers ample opportunity to grow—and to retract—with relatively low penalty and high potential yield.

Use of In-House Resources

While many successful implementations have been provided by consultants brought in for a specific purpose, most of the shell-based implementations and a surprisingly large number of the high end systems were developed primarily with in-house resources. Du Pont's strategy is to use in-house resources almost exclusively in their notably successful shell-based approach.

In amassing a bottom line contribution of close to $50 million annually for ES development, du Pont actually exports the ES technology to individual departments, encouraging them to develop focused, high-payoff systems in volume. Thousands have been developed in this way and the average cost is in the range of $10–20,000 with typical ROI in the range of 5–1 to 20–1 (Mahler 1987).

Stressing the PC as a Developmental Platform

With the distinctions between PC's and work stations diminishing (Ruth) it is an added benefit of the shell-based ES that they are typically delivered for use in microcomputers. Since a powerful work station or a very "high end" PC can be purchased for $10,000 or less, the use of these platforms greatly reduces the entry cost of ES development. Hence, the typical user and developer can begin experiencing the system dynamics without obtaining a new, dedicated machine.

Summary

This chapter offers a cost-based view of the implementation process for knowledge based systems. The unit cost and the total cost of high end ES are, in the million dollar range for large, complex systems. An alternative, offered only for the numerous ES environments that are smaller in scope, is the use of shell-based tools, which have unit costs in the range of ten per cent of the high end systems. Using these shell systems offers lower cost, easier entry, ability to spread the technology across departments without major expense (as in the du Pont case) and an opportunity to gain needed experience for in-house personnel at low cost. But the ease of use and low cost of shell systems should not encourage the indiscriminate implementation of systems without regard to appropriate points of leverage within the organization.

References

Bobrow, D., S. Mittal, and M. Stefik. September, 1986. Expert systems: Perils and promise. *Communications of the ACM* 29(9):880–894.

Engming, L. July, 1986. Expert systems for business applications: Potentials and limitations. *Journal of Systems Management* 37(7):18–21.

Gevarter, W. May, 1987. The nature and evaluation of expert system building tools. *Computer* 20(5):24–41.

Kastner, J. et al. Winter 1986. A knowledge-based consultant for financial marketing. *AI Magazine* 7(5):71–79.

Kunz, et al. 1978. A physiological-based system for interpreting pulmonary function test results. Report HPP 78–19, Heuristic Programming Project, Stanford University, Stanford, CA.

Kupfer, A. October 12, 1987. Now, live experts on a floppy disk. *Fortune* 116(8):69–82.

Lemmon, H. July 4, 1986. COMAX: An expert system for cotton crop management. *Science* 233:29–33.

Leonard-Barton, D. and J. Sviokla. March–April, 1988. Putting expert systems to work. *Harvard Business Review* 66(2):91–98.

Mahler, E. Spring 1986. *The Business Needs Approach.* Lecture on Texas Instruments' Second AI Satellite Seminar – Knowledge-Based Systems. Texas Instruments Company, Dallas. Videotape.

Mahler, Dr. Ed. November 23, 1987. Interview. *Computerworld* 21(47):S9–S11.

McGraw, K. and K. Harbison-Briggs. 1989. *Knowledge Acquisition: Principles and Guidelines.* Englewood Cliffs, NJ: Prentice Hall.

Prereau, D. 1990. *Developing and Managing Expert Systems.* Reading MA: Addison-Wesley Publishing Co.

Ruth, S. Winter 1988. Introducing expert systems in the business school: a shell game. *Interface* 9(4):42–49.

Ruth, S. and C. Carlson. 1989. Low-end expert systems in business; high yields for stand-alone applications. *Office Technology and People* 4(4):299–312.

Ruth, S. February, 1990. PC's and work stations. *IEEE Spectrum* 25–26.

Sheil, B. July–August, 1987. Thinking about artificial intelligence. *Harvard Business Review* 65(4):91–97.

Sviokla, J. 1990. An examination of the impact of expert systems on the firm: the case of XCON. *MIS Quarterly* 14(2):127–140.

3

Roles for Expert Systems in Agricultural Development

Victor S. Doherty

Introduction

This chapter suggests ways in which expert systems can contribute to solving problems faced by farmers in developing countries. The analysis focuses first on the farmer's job, one of applying managerial human capital. It then relates some of the principal issues in agricultural development to particular strengths of the expert systems approach. The result is a set of specific proposals for expert system development strategies, aimed at farmers on the one hand and scientists on the other.

Farmers and Systems

Farmers make their agricultural decisions within a context: past events, present conditions, and expectations for the future. Past provides the *status quo ante*, and materials to work with. Future presents a combination of risk and uncertainty. For the farmer, success depends upon the ability to choose the right course of action in the present. Such a choice is only possible if it is based on knowledge of all these contextual factors, along with their interactional properties. In a word, farming is a process of informed management.

Change is complicated by the fact that a farm is a functional system: each factor of production that is to be manipulated depends for its effect upon the presence and interactions of all the other factors. In order to act with informed knowledge to develop a farm, one must know the system, and the limits of effective variation of each factor within it.

Possession of an intimate, working knowledge of limits and interactions implies that one can recognize the point at which a change, such as a new level of irrigation, moves the farm from one identifiable, potentially stable system to a new one. The new system may or may not require large adjustments, such as rescheduling and major increases in fertilizer application. It may or may not be suited to help farmers in reaching goals that they sought to reach by using the old system. If the new system is not suited to the old goals, it may or may not compensate for this by its applicability to striving after desirable new goals.

Because of this complexity, farmers' adoption of new varieties of established crops is not an automatic process. A crop variety's biologically desirable new characteristics may be linked with a number of new requirements that mean the establishment of a new cropping system. Those farmers who are unable to change, whether because of lack of money, or because of unsuitable soil, or because they have other priorities, may be fully appreciative of the new variety, and equally fully adamant in their refusal to plant it.

Development Planning and the Recognition of Systems

These sorts of limitation were recognized and were addressed explicitly in India's programs, during the 1960s and 70s, to reach foodgrain self-sufficiency (Ford Foundation 1959). The key to success was not defined simply as adoption of high-yielding rice or wheat, but rather as adoption of the new variety as part of what was called a package of practices. The availability of fertilizer, and the availability of irrigation water at certain levels and at certain times, were considered as indispensable as the new, fertilizer-responsive varieties themselves. The main, early program concentration was in districts where necessary irrigation facilities and support for fertilizer distribution were already present or could be provided with the greatest speed. India's success in meeting and overcoming a national food emergency is commonly credited not only to the availability of high-yielding varieties, but also to the recognition of this systematic interdependence, and to action taken in accordance with it.

A common criticism of these high-yielding variety programs has been that in local, economic structural terms they were of most benefit to the largest farmers. The criticism hits the mark in that rich farmers who were already heavily engaged in large-scale production for the market were well served. In order to meet the threatened food deficit, the programs were concentrated primarily on production and thus they were extremely well pre-adapted for this kind of market operation. All rich farmers did not have a managerial, large-scale market production focus, however. Some were bypassed, and production on their land stagnated in relative terms. At the same time, the power of the middle class and of small farmers' and tenants' votes was not in the end ignored or turned aside: in several areas of India today, a farmer with assured control of even one irrigated hectare of land can be relatively economically secure, supported through fertilizer loans, seed distribution, and crop procurement programs.

The fact that participation by all has hardly been achieved as yet, even in such high-production areas as the rice-growing delta regions of the south or the sugar cooperative areas of Maharashtra, does not mean that wider participation is not possible. It is possible in these and other regions, in India and elsewhere. The goal should be to define the systems that need to be explored and supported to increase socioeconomic breadth of participation: to find the new packages of practices, for new environments and for new groups of farmers.

The Costs of Uniformity

A widening strategy such as this means greater complexity and greater local variation of systems. There are several reasons why complexity is desirable; not the least of these is employment. Greater uniformity can lead quickly to situations in which individual producers find it attractive to make large-scale substitution of machine power for human labor. Yet in much of the developing world, the proportion of the working population in the agricultural sector is falling only slowly, and the absolute numbers represented by this proportion are rising. Hayami and Kikuchi (1981) describe a growing stratification in some major Asian agricultural areas, and their observations might be repeated elsewhere with similar results.

Simultaneously, the technology of manufacturing is changing radically with the new capabilities available for computerization of control, design, and work (Jaikumar 1988). It is not at all clear that today's developing countries will follow the same paths to development that have been followed by others, even those whose major development has occurred in the recent past. Crucial parts of a common, working definition of economic development are that it means the growth of greater complexity, greater differentiation, and greater market participation. Such development is served by making the most of a potential for agricultural diversity, and by basing change on technological innovation that builds on human capital instead of substituting for it.

More uniformity in agriculture also implies extra expenditure for maintenance, expenditure that could be directed usefully elsewhere if uniformity were not the goal. A uniform system must be established and must be kept up in the face of whatever natural characteristics of climate and soil stand in its way. In the case of irrigation, for example, the great river basins and other areas where irrigation is in many ways a logical extension of natural systems are one thing; irrigation for most of the agricultural uplands of the world is quite another. Uniformity is not always or most often attainable at a cost that can be paid, yet with rising world population levels we will need all the production that we can get from diverse environments.

Finally, even if they can be maintained, extensive uniform systems are relatively more susceptible to disruption. When a single factor is changed, as with the advent of a disease or pest that is particularly well-adapted to a particular system, it is the extensive, uniform system that is most at risk of epidemic and catastrophe.

Farmers Are Able to Use New Information Technology

Farmers are not always aware of the latest scientific developments, but they know their own soils and climates, and they seek new factors of production (Schultz 1964). If the right knowledge about new technological possibilities is made available to them in the right way, they will know where in their existing systems the new technology fits, and they will know if it implies the development of an entirely new farming system. They will also be able to tell whether a proposed change is a logical development from existing, local environmental conditions, or if instead it is poorly conceived for the particular context. The more farmers one can work with, the more participants and willing collaborators one will have in a process of technological change.

We need not worry that farmers are not ready for the application of new information technology. Poverty and ignorance can impose extreme handicaps on large portions of rural populations. At the same time, all farmers must be observant and opportunistic in order to survive. There are particularly inquiring and perceptive farmers in all economic strata and in most villages. In order to make the most of this kind of human capital, the goal should be to find ways to make knowledge and new options more widely available.

Literacy in itself is not a problem: rural education has made strides enough in most developing countries to make it likely that even if farmers cannot read and write their own language they have easy access to a family member or a member of a neighbor's family who can do so. Again thanks to educational development, we can expect that in the early stages of an expert systems effort, when different strategies and software systems are being tried, even the use of programs whose interface is in a link language such as English or French need not be a drawback.

Hardware Availability

Before considering how expert systems might fit into the solution of agricultural development problems, we need to look at the likelihood of hardware and software availability to support such systems. The brief answer is that it appears improbable that hardware and software need be drawbacks in the application of expert systems to agriculture in developing countries.

We may deal first with hardware. Machine size and performance are certainly not problems. By now the trends to smaller machine size, to portability, and to connectivity have proceeded so far in the developed countries that desktop machines or portables are becoming the basic units in many contexts, rather than mainframes or even mini-computers. Processing and storage capacity are increasing at the same time, and the variety of technologies available for these functions is also increasing. Large computers now are for applications such as accounting functions for banks or large firms, or for modelling very large natural systems in scientific research. Basic analysis and ordinary modelling now are usually carried out on desktop or portable machines, under the control of individual users.

Microcomputers are still expensive in the developing countries, but even this situation is likely to change. Much of the manufacturing for the developed countries is carried out in the developing countries, and many of the latter have growing computer industries of their own. As these growing industries continue to advance, they will need markets, so that even if computers from the U.S. or Japan or Europe are barred by tariffs there will be pressure within the developing countries themselves for computer prices to fall so that markets may expand.

Portability and need for electricity supply are also changing, in directions that will make computing by developing country farmers more possible. In early 1989, portable computers of a size and weight that allowed them to be called pocket computers, yet with standard keyboard layouts, were being introduced in the United States. By 1990, many portables had capabilities the same or better than many desktop machines. In the meantime, storage and battery technologies were being greatly improved.

Software Availability

The purpose of creating an expert system is to represent and organize knowledge in a particular area, in such a way that it can be used to solve problems whether or not the user qualifies personally as an expert in the field. Expert systems also provide an efficient way to benefit from machine computation when ordinary algorithmic programming would be inefficient because too many possible interrelations and paths of reasoning exist to be examined exhaustively (Winston 1984). Control is not relinquished to the machine: successful systems usually are capable of displaying, upon demand by the user, summaries of the line of reasoning used and explanations of the decisions made during this reasoning (Duda and Shortliffe 1983). The user sets the task, as well. Initial hypotheses for testing are provided by the user in a backward-chaining system, which checks to see whether the preconditions for the hypothesis have been met (Duda and Shortliffe 1983). Similarly, the user provides the initial data that a forward-reasoning system uses to predict a future state (Lemmon 1986). Users also determine and describe the cases which are used inductively by some system building tools to construct possible lines of reasoning, and users approve or disapprove the results (Michalsky et al. 1985).

The simpler expert systems development tools available today (Harmon et al. 1988) are designed for user building and maintenance of these kinds of decision and analysis system. This means, for our purposes here, that agricultural knowledge, not the technician's computer knowledge, will be the critical variable in the end. As with hardware, the trends are to decentralization and user-oriented flexibility.

The fact that the inference engine for a structured, rule-based system can be separated from its particular content means that a system which has been successful in one field can be loaded with rules for quite a different field; success in this new field will then depend mainly on the quality of the new set of rules. This gives the structured, rule-based expert system approach something of the status of a technology rapidly approaching maturity, and thus the capability of being commoditized: one popular rule-based expert systems development tool sold today for general business use was first based on the inference engine from Stanford University's MYCIN. Software manufacturers find themselves competing as much in terms of the connections they offer to other programs, as they compete in terms of the artificial intelligence capabilities of their inference engines. One pays more for better graphics, or for better access to outside databases or spreadsheets or other packages, in order to have additional or more powerful ways to provide the answers that different steps in the expert system require.

New expert systems in industry are now as likely to be microcomputer- as mainframe-based. Several microcomputer-based expert systems development tools have been widely accepted by now, and the low-end versions of these tools sold in 1989 for one or two hundred dollars. As far as cost in the developing countries is concerned, work to expand the range and power of software applications there would reinforce a market which has already begun to grow, with the result that new packages would be developed at prices to suit that market.

Agricultural Expert Systems in the Developed Countries

To date the applications of expert systems in the developed countries exhibit two main characteristics. The first is *flexibility*. A wide range of potential applications for expert systems was envisioned at an early stage in the development of the field; such a range has now been realized in part, and is expanding rapidly. The second is that the warnings of the basic textbooks in the field (Winston 1984 e.g.), which caution that expert systems should be in *carefully limited domains*, continue to be well-taken.

Flexibility is increased by the fact that the types and combinations of reasoning and knowledge representation supported by microcomputer-based expert systems development tools continue to multiply. The need to restrict system domains, however, continues. Existing expert systems applications in agriculture in the developed countries exemplify these same characteristics.

There are several main trends in expert systems work in agriculture in the United States. These cover a number of possibilities for such applications: control of machinery such as grain driers (Peart et al. 1986); plant disease diagnosis (Michalsky et al. 1985); crop modelling and prediction of fertilizer and irrigation needs (Lemmon 1986). Crop modelling and growing condition modelling are the focus of much interest, and we can expect that they will be expanded greatly in a few more years. Work is underway in many places (Botcher 1985 e.g.) to use remote hookups and modelling with distributed data; this means that extension workers and farmers will be able to connect directly and on a daily or even automatic, machine-monitored basis to large-scale systems with sophisticated models and databases. The newest technology is sought to be used as soon as it is available, and the overall trend is to regionally based modelling for complex environmental and crop systems.

It would be difficult just at present for the farmers or the agricultural support services of most developing countries to meet the conditions for some of these approaches. The data and the communications infrastructure that would be required for successful programs using remote hookups and distributed data are not yet widely available. Most farmers' capitalization and current use of technology do not even approach that of the developed country farmers who use expert systems to control machinery. Nevertheless, even in developed country agriculture much work continues to be based on structured, rule-based expert systems for such tasks as disease or pest diagnosis, and this circumstance points the way to applications which would be useful and successful under developing country conditions.

Rule-based Systems for Developing Country Farmers

Structured, rule-based systems focused on diagnosis and identification are the type most widely and most immediately useful for farmers in developing countries. Extension services, no matter how hard they try, cannot do the full job by means of direct contact. Expert systems can provide precision, replicability, and reach, to reinforce and augment extension efforts. Such systems can be queried and cross-questioned about the reasons for their recommendations. New crops can be explained in full, and the results of testing on research stations and in farmers' fields can be integrated to provide a sense of crop behavior under environmental variation. Graphics can be included to simplify interaction with the decision tree itself, as well as in the on-screen responses to "how?" and "why?"

Complex but crucial topics, such as the identification of deficiency or pathogen effects under different conditions, can be dealt with. More farmers in more different situations would be able to see the same, complete information. They would have a chance, by querying the system and by working with it, to learn enough to integrate its information with their own experience. There would be a greater chance for innovations to reach all the areas to which they are suited, and to do so on the basis of full rather than partial information. There would also be a better foundation for farmer creativity and experimentation.

A campaign to make use of this potential should begin from several strong baselines, including a survey (likely to be surprising in its positive results) of existing technological penetration to the countryside. Pilot programs could use a variety of technologies, but the choice of computer, of storage medium, of interface language, and so on should take careful account of existing situations and of how these situations appear likely to change in a particular country: whatever the current situation, it is certain to change. The beginning should be focused on problems that farmers already perceive: farmers' priorities for new possibilities, and their major questions about their existing technologies and farming systems, should be sampled carefully before topics are decided on and a pilot program is begun. Consideration should be made of the best organizational way to begin: anything from a reading-room approach, to lending libraries, to distribution or sale of diskettes, to some other approach might prove most useful, for different reasons.

Considerable resources already exist for the preparation of pilot expert systems. Agricultural research organizations have produced a variety of reports over the years that are admirably suited to summarization in expert systems. Many of these reports already contain graphics which could be adapted easily for inclusion. A widely known example of a work whose text, and graphics, and layout are almost immediately adaptable in this fashion is *A Farmer's Primer on Growing Rice* (Vergara 1979). A crucial difference between presentation of this type of material in printed form and in a computerized expert system is that the expert system, while remaining as exact as the printed form, can be both interactive and self-summarizing. Farmers' criticisms and suggestions will be particularly important in the early stages of an expert systems development effort, as a guide to how these interactions and summarizations are requested and are received. Particular technologies exist already that hold considerable promise for this stage when group observation and comment will be important: they include CD-ROM and videodisk. Although electricity requirements are greater, the ability to command larger databases and the potential to store libraries of different programs for testing with groups of farmers might make CD technology especially attractive for efforts focused on initial introduction and farmer feedback.

Computing and Analysis by Farmers

A major result of an information dissemination campaign based on expert systems would be to acquaint more farmers with computers and their capabilities. This would be so whether the focus were crops, or pest identification, or some other priority. The presentation of data along with the reasons behind its use would be a powerful stimulus causing more farmers to inquire about their own land and cropping practices in a more systematic way. We can expect that as familiarity with computers and their capabilities expands, there would be a steady increase of interest in using these machines for individual record-keeping and analysis. At some point in such a process, simple spreadsheet/database programs would begin to come into their own, and we should expect that some farmers would even wish to use expert systems tools to build their own management models.

The size and scope of most small- and medium-sized farm operations in developing countries suggest that this record-keeping and analysis need not be extensive in the beginning. However, its scope and importance can be expected to grow along with growth in agricultural flexibility, complexity, and profitability. Because this kind of computing should be expected to become more widespread in the future, any early uses should be examined to see what they imply for a later date. At first, however, information dissemination is likely to be of most immediate benefit to farmers and of most interest to them.

Modelling Farmers' Knowledge Systems and Behavior

A different strategy could be followed at the same time a program of information dissemination was underway; this strategy would be focused on description and investigation. The call recurs continually for scientists to know more about farmers' real, enterprise- and field-level conditions; the very regularity with which the call is reissued, however, suggests that existing means of providing such knowledge are insufficient. On-farm testing, for example, is an integral part of many agricultural research programs. There is a continual tension, however, between the methods and goals of the farmer on the one hand, and the methods and goals of the scientist on the other. This tension exists even in those cases where research institution mandates are written to emphasize flexibility, and benefit to the small farmer, over such measures as gross production per hectare. The crux of the problem appears to be one of control: in an on-farm situation it is extremely difficult to control variables, and this lack of control means that the scientist's findings will be open to much more question than they would be if they came from a research station situation. Since professional advancement depends upon published results from controlled experiments, there is continual pressure to emphasize activities in the research station environment. Even when a farm-level trial is conducted it can be so heavily protected, from the ordinary villager's point of view, that the farm trial and a research station trial differ only in the two soil types involved.

There is nothing wrong with this ordinary focus, in principle. Precise experiments, which can be replicated, are key factors in expanding our scientific knowledge about agriculture and crops; this knowledge in turn must form a basis for suggesting effective development interventions. What we need is a way to increase scientific knowledge, while being sure that it is increased in a way that tells us more about critical factors in the farmer's everyday world.

Modelling farmer behavior would be an effective method in the pursuit of this goal. It would have the particular advantage of making farmers' strategies and rules of thumb explicit. It would require the modeller to interview farmers intensively, to test their strategies and rules for logicality, and to dig for information about how rules and strategies are interrelated and why. The result would be a dynamic picture of farmers' knowledge, in which both the accomplishments and the gaps would be clear, and in which personal goals and social organizational or institutional influences could be understood.

The work of Gladwin (1980, 1988) and others on farmers' decision-making would be very useful in suggesting directions that such an effort could take, and in providing a preview of the sorts of decision structure that are likely to be found. Expert systems like those discussed in the preceding sections need to follow the logic of the natural biological, soil, and climatic situations they model. Systems focused on farmers' behavior, however, are a different thing, and work such as Gladwin's can help in suggesting why the two types of decision tree differ or are similar.

Small-scale, rule-based expert system tools would be useful in this kind of modelling. The potential scope of the modelling effort involved here, however, suggests the need for relatively powerful inductive packages (Michalsky et al. 1985) that determine their own rules from multiple cases entered in a database. Instead of requiring the user to enter rules explicitly, such systems induce them from the structure of the database and the values of different variables for the cases entered there. It would be particularly interesting, from a development research point of view, to compare the results of expert systems based on interviews and those induced from a database of cases. Differences in priorities and perceptions could be made explicit, and related to substantive production and marketing procedures.

Developing these sorts of expert system would mean that scientists would have a common reference point for interdisciplinary communication; this communication factor alone could be important in raising the research productivity of all of the fields touched on by a successful model. The common reference could be updated and expanded on the basis of new farm-level information and on the basis of experiments. Particularly importantly from the scientist's point of view, a system of this kind would provide a way to test ideas before committing resources to a farmer's field experiment. It would sharpen the scientist's sense of dynamics and critical variables in the farmer's situation. By improving the scientist's view of the farmers world, it would bring the scientist and the farmer closer together.

References

Botcher, A.B. June 23–26, 1985. Florida's experience with computers in extension. Paper no. 85-5020, presented at the Summer Meeting, American Society of Agricultural Engineers. East Lansing: Michigan State University.

Duda, R.O., and E.H. Shortliffe. April 15, 1983. Expert systems research. *Science* 220:261–268.

Ford Foundation, Agricultural Production Team. 1959. Report on India's Food Crisis and Steps to Meet It. New Delhi: Ministry of Food and Agriculture and Ministry of Community Development, Government of India.

Gladwin, C. 1980. A theory of real-life choice: applications to agricultural decisions. In *Agricultural Decision Making*, ed. P. Barlett. New York: Academic Press.

Gladwin, C. February 11–15, 1988. Minimum data sets for cross-cultural decision modelling. Paper presented at a symposium on The Study of Agrarian Systems: Standardizing Measurement and Minimum Data Sets, Annual Meeting of the American Association for the Advancement of Science.

Harmon, P., R. Maus, and W. Morrissey. 1988. *Expert Systems Tools and Applications*. New York: John Wiley Sons, Inc.

Hayami, Y., and M. Kikuchi. 1981 and 1982. *Asian Village Economy at the Crossroads*. Tokyo and Baltimore: University of Tokyo Press and Johns Hopkins University Press.

Jaikumar, R. 1988. From filing and fitting to flexible manufacturing: A study in the evolution of process control. Working Paper. Boston: Division of Research, Harvard Business School.

Lemmon, H. 4 July, 1986. COMAX: An expert system for cotton crop management. *Science* 233: 29–33.

Michalsky, R.S., J.H. Davis, V.S. Bisht, and J.B. Sinclair. 1985. PLANT/ds: An expert consulting system for the diagnosis of soybean diseases. In *Progress in Artificial Intelligence*, ed. L. Steels and J.A. Campbell. Chichester, West Sussex, England; and New York: E. Horwood, and Halstead Press.

Peart, R.M., F.S. Zazueta, P. Jones, J.W. Jones, and J.W. Mishoe. May–June, 1986. Expert systems take on three tough agricultural tasks. *Agricultural Engineering*: 8–10.

Schultz, T.W. 1964. *Transforming Traditional Agriculture*. Chicago: University of
    Chicago Press.
Vergara, B.S. 1979. *A Farmer's Primer on Growing Rice*. Los Baños: The
    International Rice Research Institute.
Winston, P.H. 1984. *Artificial Intelligence*. 2nd ed. Reading, MA: Addison-Wesley
    Publishing Co.

4

Expert Systems for Agrotechnology Transfer

James W. Jones

Agrotechnology

Agriculture is defined by Webster as "the science or art of cultivating the soil, producing crops, raising livestock, and in varying degrees the preparation of these products for man's use and their disposal (as by marketing)". Agrotechnology is a term used to describe the totality of the means used to perform agricultural activities and produce food and fiber for human sustenance and comfort. Agrotechnology, therefore, includes the genetic material (seeds, breeding stock), physical and chemical resources such as water, fertilizer, soil, pesticides), equipment (hoe, disks, planters), and knowledge of how to use these resources to produce agricultural products in a given environment. Agrotechnology transfer can be thought of as the transfer of the physical and biological components from the site of origin to new areas. For example, new wheat varieties developed in one country may be transferred to another country by sending seed to be planted. Or machines could be sent from places where they were developed to other locations to cultivate the same crop. The transfer of physical or chemical components is not enough, and this has been shown many times when innovations that work well in one area may not work at all in another.

Another way of viewing agrotechnology transfer is the transfer of knowledge on how to produce a crop or livestock and how to match production needs to local soil, climate, and socioeconomic conditions. This transfer of agrotechnology may occur across continents, between countries, or between states in the U.S. or between research institutions and farmers in a given production region.

Some Characteristics of Agricultural Knowledge

Agricultural production systems are very complex. Growth and production of biological organisms are affected considerably by soil and weather conditions which vary greatly over space and time. Except for some greenhouse production systems, man can only partially modify soil and environmental factors for enhancing productivity and, thus, must make decisions under considerable uncertainty. There is a vast amount of knowledge that can be used to help farmers produce desired products at acceptable levels of risk.

For example, plant breeders know which soybean varieties are adapted to an area and their resistance to various plant diseases. Soil scientists may know how much fertilizer to apply to soybeans based on soil type and management practices. Entomologists may know when insects are likely to damage a crop, the types of insects, and methods for their control. A chemical dealer may know which herbicides are useful against specific weeds in a soybean field and how much to apply and when to apply it for effective weed control. The extension engineer may know when irrigation is needed and how much water to apply and he may know how to harvest and store the grain. The economist extension specialist may know the latest methods for marketing. In any of these and other areas of crop production, unexpected problems may arise, such as yellow spots on the leaves, very heavy rainfall, cold weather, and so forth. Then what, if anything, should a farmer do?

These examples demonstrate that agricultural knowledge exists from many sources, much of which is highly specialized. Some of the knowledge is "rule-of-thumb", gained by experience, whereas other knowledge may be in the form of equations or mathematical models. Farmers in the US and other countries routinely seek out experts to help them diagnose problems and make decisions. In general, the experts may be extension specialists, researchers, fertilizer or pesticide distributors, seed salesmen, equipment dealers, private consultants, and other farmers. The main points are (1) there are real problems that have to be solved and decisions to be made in agriculture that involve considerable economic risk, (2) problems are complex and require the advice of experts in many cases, and (3) the advice of experts is routinely sought by farmers to help them in their decision making process.

In many cases, this knowledge is local and is applicable only within a limited region. In other cases, the knowledge is generic and could be applied to help solve problems in various regions or locations in the world, although with perhaps less certainty or specificity. Thus, it is useful to think of agrotechnology transfer on at least two levels of detail: local and remote. A local knowledge base is likely to be more specific and based to a large extent on experience and on acquired local knowledge. When transferring agrotechnology to remote areas where there is no local experience with that crop/variety/practice, there may be somewhat lower levels of confidence in the solutions to problems or decisions until they are tried. A major benefit of agrotechnology transfer to new areas using ES is that the number of experiments in the new area could be reduced considerably by knowledgeably eliminating those production practices that have high likelihood of failure—for any of many reasons.

Developing vs. Developed Countries

In the US and other developed countries, one may think that agriculture is sufficiently well developed that there is no need of such sophisticated tools as ES. We have many outstanding agricultural research institutions and a public-funded, nationwide extension service with offices in all counties for transferring knowledge to farmers. There are also various other public and private organizations for providing farmers with information for decisions. Why are ES needed at all? On one hand, US farmers operate in world markets where competition is stiff. Profit margins are low and farmers must manage their farms to maintain adequate profits at acceptable risks. This can only be done through good decisions based on knowledge of the factors involved. On the other hand, sources of knowledge in US agriculture are diverse. The local extension agent cannot be an expert in all areas of production, for all crops and livestock. The lack of availability of a limited number of experts may result in farmers making poor decisions that cost considerable money. ES have potential for making knowledge more widely available on a more timely basis. Also, experts retire and move to other positions. ES can help preserve expertise when such transitions occur, and help train replacements. I have also found that ES can help tailor recommendations to specific conditions as opposed to general printed recommendations that county extension agents usually have available. Finally, ES can help in the adoption of a new technology through education of farmers and advisors. I have concluded that ES should become a major tool of extension agents who are in direct contact with farmers. The extension specialist's role should include the development and testing of ES to transfer as much of the knowledge that they can represent to extension agents, consultants and farmers.

The situation in developing countries is not too different, except that problem solutions and recommendations may vary considerably from those in more developed countries. Furthermore, few (if any) farmers are expected to use any ES directly. The major use of ES in agrotechnology transfer in developing countries will be for educating research and extension personnel who can then provide more knowledgeable advice to local farmers. In some countries, they may play a role in land use planning where new agricultural developments will be planned, based on expert knowledge of crops, soil, weather, socio-economic conditions, and government goals and policies.

Examples of Existing ES in Agriculture

The current availability of ES for agrotechnology transfer is very small relative to the potential for useful systems. Development of ES in agriculture is in its infancy. However, several ES have been developed ranging in applications from those for diagnosing problems to those that prescribe management actions, help farmers plan production, market their crops and livestock, and control greenhouses or irrigation equipment. In these systems, various ES methodologies have been used. Production rule and framebased ES have been developed. Expert databases with a natural language interface are being developed. ES have been linked with simulation models and some work on knowledge based simulation in agriculture has been reported. Finally, there are several current efforts to develop integrated decision support systems for agrotechnology transfer consisting of databases, simulation models, and ES integrated together. A few of the existing agricultural expert systems will be noted below.

Examples of ES using the Production Rule Approach

One of the major application areas of ES has been in diagnosing problems. PLANT/ds was one of the first ES in agriculture (Michalski et al. 1982). It was developed to help farmers and extension agents correctly diagnose diseases of soybean. A similar ES called TOM was developed in France to diagnose diseases of tomato plants. A third diagnostic ES called CHAMBER was developed by Jones and Haldeman (1986) to diagnose problems in the control of a sophisticated crop research facility. CHAMBER was developed on a microcomputer using the INSIGHT shell (Level V 1987). The ACID4 ES (Yost et al. 1986 and Chapter 9) was developed to diagnose possible pH problems in soils of the tropics and to recommend lime applications for producing various crops. This ES contains crop and soil data bases, uses production rules, and is implemented on microcomputers. Beck et al. (1987) developed an ES called SOYBUG on a microcomputer using INSIGHT2+ (Level V 1987) for determining the need to apply insecticide to soybean in Florida and recommending the insecticide to use depending on the density and types of insects in the field. W. G. Boggess (Personal Communication, University of Florida 1988) developed an ES using production rules to diagnose the financial stability of a farm and to present information to the user about why the conclusion was reached. Jacobson et al. (1987) developed a real-time expert system to integrate expert knowledge with sensors and control devices in a greenhouse to supervise the control of environmental conditions inside the greenhouse. In this system, the rulebased ES interacts with a dynamic database and creates new control actions for heaters, vents, and carbon dioxide injectors every few minutes. A similar real-time expert system for controlling misters in greenhouses was reported by Jones et al. (1986). One characteristic of most of these examples is that they were developed and implemented on microcomputers. This has and will continue to be very important in the subsequent distribution and use of these systems by agricultural users who are scattered widely.

Production Rules Linked with Simulation Models

During the last two decades, agricultural scientists have made considerable progress in the development of simulation models that describe the behavior of crop and animal systems in response to various environmental and management inputs. For example, crop simulation models are now available for predicting growth and yield in response to soil water availability, temperature, solar radiation, photoperiod, row spacing, planting date, and in some cases soil nutrient availability. These models have been used successfully by researchers who have demonstrated their validity under certain conditions. However, the use of these models by others for decision making has not occurred on a broad front. The major reason for this lack of use is the fact that the models do not account for every possible stress that may occur in the real world (such as diseases, weeds, etc.) and as such, the model results must be interpreted in the context of other factors. Model developers who use the models are careful to interpret results with these limitations in mind. I have concluded that the capabilities that ES provide for reasoning complement the capabilities that simulations provide and that the merging of the two concepts can capture the best of both methodologies for agrotechnology transfer (Jones, Jones, and Everett 1987). In agriculture, this merging is occurring on two fronts.

First, traditional simulations are being interfaced with expert systems in what can be termed a hybrid approach. The interface allows information exchanges which enhance both the simulation and expert system. The second way, in which the approach to modeling is fundamentally changed, is through knowledge-based simulation. In knowledge-based simulation, automated reasoning about models allows model components to be selected and combined to achieve a goal or design. In addition, a uniform representation permits integration of databases and other applications along with simulations and expert systems. Most attempts to combine simulation and AI techniques in agriculture use the hybrid approach. However, a recent emphasis on integrating a variety of software tools for whole farm decision support are heading in the direction of knowledge-based simulation.

In the hybrid approach, an ES may select alternatives for simulation, interpret model outputs and suggest action, provide missing components for simulation, select an appropriate model, or it may perform combinations of these tasks in solving a problem. Most of the expert system-simulation combinations in agriculture are organized in a similar fashion. The expert system component defines the scope of the overall problem, the shell software interacts with the user to obtain some needed inputs, and other information is obtained from a simulation model or perhaps from a database.

In spite of attempts to make the simulation models more realistic and comprehensive, they are still only approximations of the real system. One approach is to include in a knowledge base relevant considerations not in the simulation model as well as knowledge on selecting improved alternatives. This concept was used by COMAX (Lemmon 1986) to provide a within- season decision aid to cotton farmers. A cotton growth model is used to compare various decisions concerning fertilizer, irrigation, and harvesting. The expert system selects the combinations of alternatives to compare and interprets simulated outputs to provide users with improved cotton crop management decisions.

Many simulations produce a large amount of output that is difficult to interpret except by those most familiar with the simulation. This is true both in continuous agricultural models, such as crop growth, and in discrete, stochastic models. Khuri (1987) developed a discrete, stochastic model for describing the utilization of labor and machinery in citrus harvest operations. The goal of that work was to provide managers of citrus harvest operations a tool for better selecting the combinations of equipment and labor. The simulations produce information on equipment utilization, idle time, and harvest volume for a particular combination. However, citrus harvest operators could not effectively interpret the output, so a knowledge base was developed for interpreting simulated results for users and suggesting a course of action. The resulting system, CHESS (Citrus Harvest Expert Simulation System) now interprets the critical indicators of system performance and provides recommendations on how the harvest operation can be improved.

Simulation models have limitations in describing real-world system behavior. For example, crop system models have typically included soil, water, climate, and crop growth components while omitting most biological organisms. In some cases, insect models have been included, but perhaps only for one of the principal insect pests that would likely occur in a real field. Furthermore, quantitative models of insect pests are difficult to apply in a real system because of the data required to run them and the inherent variability in population dynamic processes. In contrast, entomology experts may be able to predict fairly accurately the pest populations in a crop, based on field sampling and past experience. McClendon et al. (1987) reported a soybean pest management expert system (SMARTSOY) which uses scouting data and expert knowledge to project insect populations for the next week in a crop of soybeans. Estimates of the damage potential by all insects are fed into the SOYGRO crop growth model (Jones et al. 1987) to estimate pessimistic, optimistic, and expected yield loss that would occur in the next week if control action is not taken. Depending on the yield loss and the grower's sensitivity to risk, recommendations are then made by the expert system as to the type of insecticide to apply for the given complex of insects. The target audience for this system is extension agents and pest consultants. The expert system is written in INSIGHT2 + and the simulation model in FORTRAN, both operating on a microcomputer.

Frame-Based ES

POMME (Roach et al. 1986) is a consultation system for apple orchards. It utilizes a knowledge representation approach in which rules and frames are integrated. Rules are represented as frames with the antecedent and consequence parts of the rules treated as slots. Users can obtain advice on a number of topics, such as pest control recommendations. Computer simulations can also be integrated within the frame-based structure of POMME. The system designed for nursery production of seedlings presented by Rafae, Warkentin and Ruth in Chapter 7 represents another example of a system integrating both rules and frames.

Beck (1988) reported on a prototype frame-based ES for accessing a large database on pesticides. The basis for the frame structure was derived from the KANDOR (Patel-Schneider 1984) system. A natural language interface was developed to allow users easily to obtain the information needed for a particular problem.

Decision Support Systems

Decision support system (DSS) is the generic term applied to a system which acts as an aid to decision makers. Since the DSS may be called upon to answer *ad hoc* questions, it must contain a broad and detailed knowledge base. A combination of expert systems, databases, and simulations provide a good mixture of techniques in a DSS. It is important to consider how all these components will be coupled and integrated. Ideally, a knowledge base would be capable of reasoning about its components and select the proper components based on an inferencing mechanism. More common is a simple menu selection approach in which the user must manually search the DSS for appropriate tools.

There has been much recent interest in decision support systems for agriculture. The following examples illustrate various approaches.

COTFLEX (Stone et al. 1987) is a decision support system for farm-level decisions about individual crops. A frame-based representation is at the core of the system. Information on the current status of particular fields is stored using frames. Particular fields are instances of generic field frames which identify types of crops, particular crops, and particular cultivars. The slots of the frames contain attributes describing the status of each field (stage of development, pest population levels, etc.). The values in these slots can come from one of four possible sources, 1) a simulation, 2) an expert system, 3) a database or 4) from the user. When information is needed for a slot, a decision is made about which source is best. COTFLEX also includes a time representation for seasonal operations to be performed for specific crops.

CALEX (Plant and Wilson 1987) is a microcomputer-based decision support system which provides access to data files, expert systems, and simulations. Access to these modules are controlled by an executive program. Users interact with the executive program which presents data collection forms, and controls access to the sub-modules. Facilities are provided by which expert systems can call simulations, and vice versa. CALEX stresses the record keeping aspects of farm management decisions. Thus, a database for storing field and weather data is central to the system.

An international team of scientists has developed a Decision Support System for Agrotechnology Transfer (DSSAT) which currently contains five crop simulation models; databases of weather, soil, and experimental crop-soil data; one expert system for recommending lime application to soil; a risk-analysis package; graphics, and several user-oriented programs (IBSNAT, 1986; Uehara, personal communication). DSSAT provides an environment in which simulations, databases, and expert systems can easily be selected, connected, and executed all on microcomputers. Users identify which components to run by using a menu selection supervisory program written in "C". The DSSAT is being used in educational workshops around the world, and by teams of scientists who evaluate alternate crop production systems in various countries and recommend policy to their respective governments. The system is being expanded to include additional simulations and expert systems.

Conclusions

There is considerable activity in developing ES for agrotechnology transfer today. The concepts being applied range from production rule systems, to frame-based systems, to integrated systems containing databases, simulations and expert systems. Some work is being initiated using natural language interfaces and object oriented Programming. Almost all of the applications in agriculture are being developed and implemented on microcomputers. This is expected to continue so that distribution and use of the systems by widely scattered users in developed countries will be possible, and so that they can be used in developing countries where microcomputers are becoming more widely available. Because of the many critical problems faced by agricultural producers and the nature of the knowledge, ES will play an increasingly important role in the representation, storage and transfer of knowledge.

References

Barrett, J., and D.D. Jones. 1989. *Knowledge Engineering in Agriculture*. ASAE Monograph No. 8. St. Joseph, MI: Amer. Soc. of Agr. Engineers.

Beck, H.W. 1988. Expert data base management. In *Knowledge Engineering in Agriculture,* ed. J. Barrett and D.D. Jones. ASAE Monograph. St. Joseph, MI: Amer. Soc. of Agr. Engineers, 83–115.

IBSNAT. 1986. Decision support system for agrotechnology transfer. *Agrotechnology Transfer* 2:1–5. International Sites Network for Agrotechnology Transfer Project. Honolulu: University of Hawaii.

Jacobson, B., J.W. Jones, P. Jones. 1987. Tomato greenhouse environment controller: Real-time expert system supervisor. ASAE Paper No. 87-5022. St. Joseph, MI: Amer. Soc. of Agr. Engr.

Jones, J.W., K.J. Boote, S.S. Jagtap, G. Hoogenboom, and G.G. Wilkerson. 1987. SOYGRO V5.4: Soybean crop growth simulation model: User's Guide. *Fl. Agr. Exp. Sta. Jour. No 8304*. Gainesville FL: Agr. Engineering Department, University of Florida.

Jones, J.W., P. Jones. and P.A. Everett. 1987. Applying agricultural models using expert systems concepts. *Trans. of the ASAE* 30:1308–1314.

Jones, P., and J. Haldeman. 1986. Management of a crop research facility with a microcomputer-based expert system. *Trans. of the ASAE* 29(1):235–242.

Jones, P., B.K. Jacobson, J.W. Jones, and J.A. Paramore. 1986. *Real-Time Greenhouse Monitoring and Control with an Expert System*. ASAE Technical Paper No. 86-4515. St. Joseph, MI: Amer. Soc. of Agr. Engr.

Khuri, Ramzi S. 1987. *An Expert System for Citrus Harvest Analysis*, Masters Thesis. Gainesville, FL: University of Florida, Department of Agricultural Engineering.

Lemmon, H.E. 1986. COMAX: an expert system for cotton crop management. *Science*. 233:29–33.

Level 5 Research, Inc. 1987. *INSIGHT2 +*. 4980 South A-1:1, Melbourne Beach, FL.

McClendon, R.W., W.D. Batchelor, and J.W. Jones. 1987. *Insect pest management with an expert system coupled crop model*. Technical Paper No. 87-4501. St. Joseph, MI: Amer. Soc. Agr. Engr.

Michalski, R.S., J.H. Davis. V.S. Bisht, and J.B. Sinclair. July, 1982. PLANT/ds. An expert system consulting system for the diagnosis of soybean diseases. *Proc. of Fifth European Conf. on Artificial Intelligence*. Orsay, France.

Patel-Schneider, P.F. December, 1984. Small can be beautiful in knowledge representation. *Proceedings of IEEE Workshop on Principles of Knowledge-Based Systems*. Denver, CO.

Plant, R.E. and L.T. Wilson. 1986. A computer based pest management aid for San Joaquin Valley cotton. *Proc. Beltwide Cotton Conf.* Memphis, TN: National Cotton Council of Amer: 22–24.

Roach, J.W., R.S. Virkar, M.J. Weaver, and C.R. Drake. 1986. POMME: an expert system to manage pests in apple orchards. *Abstracts of papers*, 152nd National Meeting. AAAS. Philadelphia.

Stone, N.D., R.E. Frisbie, J.W. Richardson, and R.N. Coulson. 1986. Integrated expert system applications for agriculture. *Proceedings of International Conference on Computers in Agricultural Extension Programs*. Lake Buena Vista, FL.

Yost, R.S., G. Uehara, M. Wade, I.P. Widjaja-Adhi, and Zhi-Cheng Li. 1986. *Expert systems in agriculture: Determining the lime recommendations for soils of the humid tropics*. Series 000. Honolulu: Hawaii Institute of Tropical Agriculture and Human Resources, University of Hawaii.

5

AGASSISTANT: A New Generation Tool for Developing Agricultural Advisory Systems

Thomas W. Fermanian and Ryszard S. Michalski

Introduction

AGASSISTANT is a new generation expert system builder for personal computers in the domain of agriculture. While it may be used to construct expert systems in any domain, its inference system was designed specifically to handle the uncertainty found in many agricultural domains. It is an extension of a conceptual predecessor, PLANT/ds, an earlier expert system for the IBM PC which was concerned with the diagnosis of soybean diseases common in Illinois. PLANT/ds was the first agricultural expert system that had the capability inductively to learn rules from examples, in addition to acquiring them directly from an expert (Michalski and Chilavsky, 1980b; Michalski et al. 1982). Unlike PLANT/ds, in which one interacts with the VAX minicomputer to build and refine the knowledge base, one need not leave the PC environment, either in creating an expert system, or in getting advice from a system that already exists. The work here is also based to a large extent on the ADVISE meta-Expert System (Michalski and Baskin 1983).

Among the many important features of AGASSISTANT are:

Multiple means of creating and refining knowledge. AGASSISTANT can receive rules directly or acquire them through inductive inference.

The authors wish to acknowledge J.C. Fech and J.E. Haley for their assistance in conducting the WEEDER evaluation study and to thank B. Katz and J. Kelly for their programming support.

Probabilistic inference is employed for handling uncertainty of data and rules.
It is of great use in agriculture, where the vagaries of nature make identification or diagnosis a probabilistic matter.

Implemented on a personal computer. This allows wide dissemination of the
program to farmers and others in need who are unlikely to have access to
larger systems due to a specially designed human-computer interface.

Menu-driven screens. The novice user can quickly come up to speed in
building expert systems due to a specially designed human-computer
interface.

An Overview of the Program

The AGASSISTANT advisory system consists of a set of modules accessible through menu-driven screens (see Figure 5.1). The Advisor module takes
as its input a compiled expert system, which it uses to create questions for the
user, as well as to give advice on the basis of the answers to these questions.
The Compiler module, in addition to parsing rules for correct syntax, creates a
more compact version of the system for faster execution. Rules may be created
by hand, or created through induction from examples. The Inference Engine
module calls the program NEWGEM (Reinke 1984) to operate on examples,
and variable definitions to produce rules. Additionally, this module may start
with existing rules and improve them with examples (incremental learning), or
optimize them according to differing criteria.

Knowledge Representation in AGASSISTANT

Knowledge is represented in AGASSISTANT primarily in the form of
attribute-based logical calculus, called VL1 (Michalski 1975). Variables in

FIGURE 5.1. An Overview of AGASSISTANT

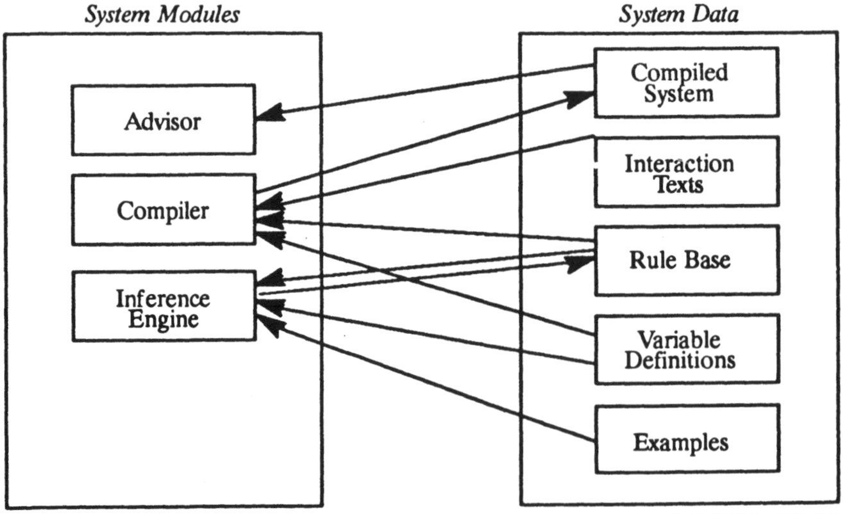

these rules may be nominal, linear, integer, or structured. The exact syntax and constraints on the rules are described in the section on knowledge acquisition. Here we will simply give the reader a feel for the way knowledge is represented so that he or she may better understand the following sections. A hypothetical rule that incorporates all of the allowed variable types appears in Figure 5.2.

The above rule contains two complexes (a conjunction of elementary conditions), the first of which consists of four conditions (also called selectors), while the second complex contains two selectors. Each selector is followed by a weight or confidence level (CL), which indicates the relative importance of the selector as a sole condition for making the decision. For example, if the only fact known is: "the crop shape is oblong," then the expert is 40% confident that the crop should be harvested. The method of combining these weights if more than one fact is known is explained in the Advisory module section. The action for this rule, namely that the crop should be harvested, depends on whether enough of the conditions are satisfied. The degree to which the set of conditions must be satisfied can be set as a system threshold. Unlike other inferencing mechanisms, this system can support a decision with only an approximate match of the evidence to the stated conditions.

The first condition will be satisfied if the shape of the crop is oblong. Since this is a structured variable, values of the variable are arranged in a hierarchy; thus this condition will be true if crop-shape takes the value of oblong, or any child of oblong. The second selector will be true if the temperature is between 65 and 75. The third selector will be fulfilled if soil-moisture is in the range of medium to high; there may be values in between medium and high (such as medium-high) which are implicitly included in this selector. The last condition will be satisfied only if the nominal variable sky receives the value sunny, out of a possible set of values including raining, cloudy, etc. The second complex of the rule, following the 'OR' consists of two selectors and says more or less that if you haven't brought in the crop by October, you should do so now if the weather is fair.

FIGURE 5.2 A Hypothetical Rule in the Knowledge Base of an Expert System

Crop should be harvested if:	CL
1. Crop-shape is oblong,	40
2. Temperature is 65 to 75,	45
3. Soil-moisture is medium to high	10
4. Sky is sunny.	5
OR	
1. Month is October,	75
2. Weather is fair.	25

A knowledge base for an expert system consists of a set of such rules. Rules may also be structured in a hierarchical fashion, i.e., a condition of one rule may be the action of another rule. The section on the exemplary expert system WEEDER outlines a small complete expert system.

Relevance to Agriculture

Many underlying principles of agricultural sciences can be experimentally measured in field experiments. The results of these experiments, however, show a normal abundance of variation in measured responses which is expected in nature. This natural variation has made the development of realistic models of agricultural systems difficult. Many assumptions are required for even the most simple crop or environmental model. Expert systems technology, for the first time, will offer a technique for working with fuzzy or uncertain knowledge.

Agricultural scientists often provide advice to agricultural managers on the basis of an evaluation of their incomplete knowledge and experience. This closely parallels the process of an expert system. Due to natural variability, agricultural knowledge bases are unstable and require continual modification to reflect current conditions or knowledge. Advisory systems which require the intervention of mainframe computers or centralized systems development cannot keep up with the rapid changes necessary. AGASSISTANT represents an expert system development environment that can be easily modified in the field. Therefore it can truly reflect any changes seen in the natural responses of the model in question.

While many agricultural production processes are inherently complex, a subgroup of production processes can be described adequately and converted to an appropriate knowledge base for use with AGASSISTANT. An example of these are pest or crop identification or pest damage diagnosis systems. In addition, simple designing of agricultural production systems would be an appropriate domain for AGASSISTANT. Often agricultural data are of a subjective, qualitative nature which might be more rapidly and thoroughly analyzed through symbolic processing techniques. AGASSISTANT represents a new technology in the form of a tool to assist agricultural scientists and managers to better interpret the observed phenomenon.

The Advisory System

While the Advisor module is most apparent to the user, at the heart of the AGASSISTANT is a flexible inference engine that runs on a compiled set of rules. The exact form of this file is described in detail in the first part of this section. The second part explains how individual rules are evaluated and how uncertainties are propagated through a hierarchical knowledge base. The third part explains the control structure of the system, i.e., the method by which the advisory system generates questions. This section explains the technical details upon which the advisory system is based.

System Representation

The rule parser takes a set of rules and a set of variable definitions and produces a file containing the compiled version of the expert system. This file consists of a cross-referenced version of the original system. Figure 5.3 shows a section of a rule base for an expert system, while Figure 5.4 illustrates the definitions of the variables involved.

The rules in the above system represent a section of a hypothetical expert system for crop management. Undoubtedly, an actual system, would contain many additional rules, (e.g. rules for fertilization, plowing, etc.) and each rule would be of greater complexity; this rule base is meant solely for illustrative purposes. The rules indicates that the crop should be harvested if the weather is good and the crop is ripe. Each of these conditions are in turn based on further conditions as indicated in the first two rules. The variables and their respective types and domains are shown in Figure 5.4. The compiled system appears in Figure 5.5.

Before explaining the desirability of converting the original system into its compiled form, the format of the file in Figure 5.5 will be explicated. The file consists first of a list of variables and information relevant to them. For example, the variable Crop-color is the fifth variable in the list (after Soilmoisture). Below the variable name is the relation used with this variable, and then the values associated with the variable obtained from the variable table. Following each value is a list of tuples, each containing four elements, *viz*, rule number, rule complex, selector within the complex, and a weight. For example, if Crop-color takes the value yellow, then selector 1 of complex 1 of rule 2 (rules are given numbers in order of appearance within the rule base) will be updated by 50% (the weights are represented as percentages with an additional digit of significance to reduce the possibility of roundoff error). Values may have more then one tuple following them if the variable value pair appears in more than one complex.

FIGURE 5.3 An Illustrative Rule Base

Crop should be harvested if:	CL
1. Crop is ripe,	60
2. Weather is good.	40
Crop is ripe if:	
1. Crop-color is yellow or green,	50
2. Crop-shape is oblong.	50
Weather is good if:	
1. Sky is sunny,	50
2. Soilmoisture is low,	50
3. Windstrength is very-mild to medium.	25

FIGURE 5.4 Variable Definitions for Illustrative System

Variable	Crop	Weather	Soilmoisture	Crop-color	Windstrength	Crop-shape	Sky
Type	*nominal*	*linear*	*linear*	*nominal*	*linear*	*structure*	*nominal*
Value 1	ripe	bad	low	yellow	very-mild	round	sunny
Value 2	unripe	fair	medium	green	mild	oval pear	cloudy
Value 3		good	high	brown	medium	oblong	rainy
Value 4				black	mediumstrong	short long	
Value 5					strong		
Value 6							
Value 7							
Value 8							

Following the variables, appear two lines of #'s, after which appear the rule actions. Each action has three lines of information. The first line is simply the action verbatim, as it appears in the rule base. The second line, possibly blank, contains the name of a text file (e.g., harvest.txt for the rule "Crop should be harvested") invoked if the rule associated with the action receives the highest confirmation at the completion of the advisory session. The third line consists of a number that indicates which variable, if any, is identical with the action, followed by a list of four-tuples, with the same ordering as described above, which list the locations of the action as conditions in other rules. For example, this line for the action "Crop-shape is oblong" begins with an 8 to indicate that this action variable, *viz.*, "crop-shape", is also the eighth variable in the variable list at the beginning of the file. Additional information indicates that the action also appears in rule 1, in the first selector of the first complex.

In the fourth and last line the order in which variables should be asked is presented. Following this is a list of arithmetic expressions found throughout the rule base, with the appropriate tuple list trailing each one. Finally, variables and questions are listed that will be asked during the advisory session when the system wishes to know the value of a variable.

One may contend that the information contained in the compiled system, with a few minor additions, is just a rehashing of the system in its original form. While this is correct, there is an important reason for representing the system in such a way. Consider what needs to be done to update a rule after a new value is associated with the variable. Suppose the system consists of *n* rules each with an average of *s* selectors. One must then search the entire rule space for the appearance of the variable-value pair, or perform *ns* searches. Additionally, in the worst possible case, one must search for the actions appearing as selectors in other rules resulting in sn^2 matches, or ns matches for each of n rules. While both of these figures are polynomial quantities, they will nevertheless prove prohibitively large for a PC system if n is large. Thus the appearance of all values, as well as the location of all actions as conditions in other rules, are cross-referenced beforehand in the compiled file, resulting in almost no search. The exact method of updating rule confidence levels, once this information is provided, is examined in the following section.

FIGURE 5.5 An Example of a Compiled Expert System

Variables		Rule Actions	
############		############	
crop		############	
is		Crop should-be harvested	
ripe	111600	harvest.txt	
############			
Weather		38	
is			
bad		Crop is ripe	
fair			
good	112399	8111	
############		125	
Soilmoisture			
is		Weather is good	
low	312250		
medium		3112	
high		467	
############		############	
Crop-color		12345678	
is		############	
yellow	211500		
green	211500	$$ Arithmetic expressions	
brown		$$ (Crop-width * Crop-length) > 12 212500	
black			
############		############	
Windstrength			
is		Variables and associated questions	
very-mild	313250		
mild	313250	Weather	What is the weather like?
medium	313250	Soilmoisture	To what extent is the soil saturated?
strong			
############		Crop-color	What color is the crop?
Sky		Windstrength	How strong is the wind?
is		Sky	What is the appearance of the sky?
cloudy			
rainy		Crop-shape	Is the shape of the crop oblong?
sunny	311500		
############		Crop-length	What is the length of the crop?
Crop			
is		Crop-width	What is the width of the crop?
ripe	212600		
############			

FIGURE 5.6 General Rule Format

Action is xxxx if:	confidence level	
1. variable1 is value1,	60	
2. variable2 is value2 or value3,	40	} Complex1
3. variable3 is value4 to value6.	20	
OR		
1. variable4 is value7,	50	} Complex2
2. variable5 is value8.	50	

Inference Mechanism

Rules have the general format shown in Figure 5.6. Since the conditions in rules are annotated by weights of confidence levels to represent strength of evidence in favor of the decision, it is not sufficient simply to invoke the standard laws of deductive inference in evaluating confidence levels of the rules. The next section describes the method for evaluating individual rules, while the one following it shows how rules are evaluated in a hierarchical system.

Individual Rule Evaluation. The complex of a given rule is easy to evaluate given the form of the compiled system. A sum is maintained for each complex of each rule, and this sum is augmented by the amount indicated by the tuple associated with the satisfied variable-value pair. The general formula for evaluating a complex is:

(1)
$$\frac{\Sigma \ (weights \ of \ satisfied \ selectors)}{\Sigma \ (all \ weights \ in \ the \ complex)}$$

For example in complex1 of the general rule in Figure 5.6., if variable1 had the value of value1, and variable 2 took the value value3, while variable3 the value value9 (not in the linear range value4) then complex one of this rule would have the value of:

(2)
$$\frac{60+40}{60+40+20}$$

or 83%. Selectors with internal disjunction (selector with a set of disjunctive values) are assumed satisfied if any of their values is the current value of the variable. Selectors with linear variables are satisfied if the variable takes on any value within that range inclusive of the end points. Finally, selectors with structured variables are true if the variable receives the indicated value or some child of that value. For example, the condition "Crop-shape is oblong" will be true if either oblong, short, or long is selected.

Evaluation of rules with multiple complexes is performed by recursively taking the probalistic sum (referred to as "psum" evaluation in ADVISE [Michalski and Baskin 1983]) of the nth complex with the psum of the first n-1 complexes. The formula for this is:

(3) psum(V(n),V(n-1),...,V(1)) =
$$V(n) + psum(V(n-1),...,V(1)) - V(n)^*psum(V(n-1),...,V(1)),$$

where $V(x)$ is the evaluated value of complex x.

Other evaluation schemes such as taking the complex with the highest value are possible but are not included in this version of the system. The virtue of the psum scheme is that it fits well to the intuitive notion that if any complex is completely satisfied, the rule also will be, a notion that the probability of two independent events occurring is the probalistic sum of the event probabilities.

A rule is considered to be satisfied if its confidence level goes above a threshold, experimentally set at 85%. Notice that if all selectors are assumed to have equal weights and the threshold is set at 100%, then the evaluation schemes described here collapse into formal logic expression.

Evaluation of Rules in a Hierarchical Rule Base. Selectors in rules which are in turn actions of another rule receive the value of the rule times the weight for that selector. Thus in Figure 5.3 if one assumes that the sky is sunny in the third rule, and that the confidence that the weather is good is 50%, the confidence that the weather is good in the first rule is 50%*40%, or 20%.

A hierarchical rule base contains a partial ordering among the rules. The rule evaluation module topologically sorts the rules according to this partial ordering before the advisory session begins, and uses the resulting order to evaluate all rules after each new value is entered. This ensures that all weights are propagated in the correct sequence. Notice, that if one assumes that all selectors of all rules receive equal weights, and if the threshold of rule firing is 100%, the inference scheme emulates a standard forward-chaining inference engine.

Control Mechanism

The Advisor module of AGASSISTANT has two control mechanisms, that is, schemes which determine the order in which questions are asked. The first method applies to rule bases which are flat, or non-hierarchical in nature, and is known as the utility scheme. The second mechanism, for hiearchical rule base, is a backtracking scheme.

Utility Control Scheme. The utility control scheme is represented in Figure 5.7. At the start of the advisory session questions are asked for variables in order of the utility of the variables. The utility measure, precalculated and found in the compiled system file, reflects the degree to which the variable will affect the confirmation level of all rules. Those variables appearing in the most places in the rules are given the highest utility. The advisory session continues until the confirmation level of any rule goes above a lower threshold, experimentally set at 15% 1–upper threshold. When this occurs, the system will focus on that rule by asking for the values of all variables relevant to the rule. This continues until the rule is rejected, or all variables for that rule are exhausted. At this point the system will focus on another rule which is above the upper threshold, or if none exists it will return to the utility measure. The entire process continues until all rules are rejected or confirmed. This may require that all variables be queried for, but usually occurs much sooner. In general, a rule is confirmed if it has a confidence level above the confirmation threshold, while it is rejected if it cannot possibly be confirmed, no matter what the values for the remaining variables are. The system makes use of a method of keeping a "negative" confidence for all rules; thus there is no need to determine dynamically whether a rule is rejected after each new answer.

Backtracking Control Scheme. The backtracking control scheme is automatically invoked if the rule base is hiearchical. It is represented in Figure 5.8. The system begins by asking questions for variables with highest utility and continues in this fashion, until the user answers "don't know" for a variable

FIGURE 5.7 Utility Control Scheme

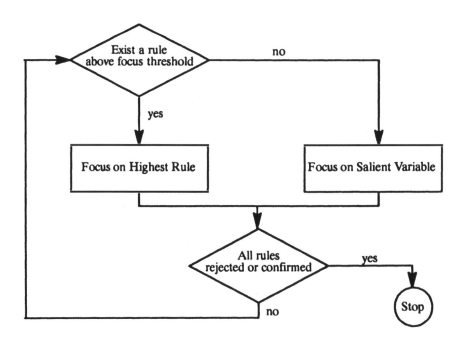

FIGURE 5.8 Backtracking Control Scheme

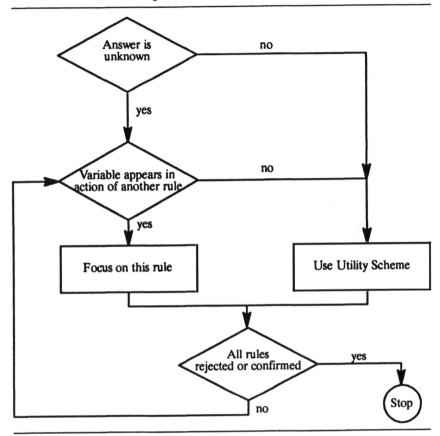

which also appears in the action of some rule in the system. For example, if the user answers "don't know" for the variable Weather in the rule base of Figure 5.3, then the system will attempt to infer the value of Weather by asking for the values of Sky, Soilmoisture, and Windstrength. The process continues recursively until the system is able to find the value for the original variable it was focusing on, in this case Weather, at which point it returns to the utility scheme. The system will continue to query the user until all rules are either rejected or confirmed.

Knowledge Acquisition Facilities

Figure 5.9 illustrates the knowledge acquisition facilities of AGASSISTANT. As stated at the outset, knowledge is represented in the system in the form of rules in the VL1 syntax. These rules can be acquired in four possible ways. They

FIGURE 5.9 Knowledge Acquisition in AGASSISTANT

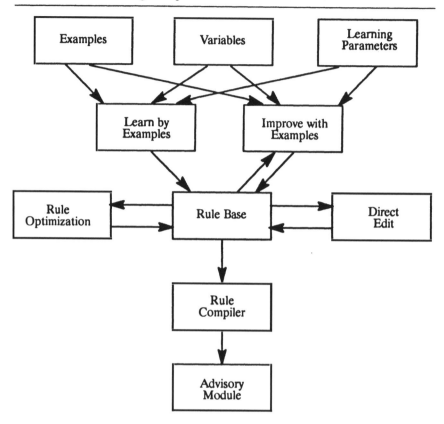

can be learned from examples, improved with examples, optimized, or edited directly. It is also possible to use these methods in combination. For example, one common procedure is to edit rules directly, and then optimize them to remove superfluous information. Once acquired, rules are compiled and serve as input to the advisor module: in effect the system's interface with the user. The various facets of knowledge acquisition are described below.

Direct Editing of Rules

One way to enter knowledge into the system is to handcraft it to the system. To use this method an expert must be able to express knowledge in the form of rules. This notoriously difficult problem is often referred to as the knowledge acquisition bottleneck (Michalski and Chilausky 1980b). For example, one may be a perfectly adequate driver and yet have difficulty expressing this knowledge in rule format. Nethertheless, there are many domains in which direct entry of rules is appropriate. In AGASSISTANT, this is accomplished in two steps. First, the relevant variables are entered into the variable editor. The variable type, and its domain, are specified within the editor. Then, upon selection of the direct edit function within AGASSISTANT's menu system, a child process will be created. This process consists of the default editor (whose name is stored in the file 'AGEDITOR.TXT') with the file to edit being concatenation of the current system name and the extension .RLE. The command processor of MSDOS will return the user to the AGASSISTANT system upon termination of the edit. The user then typically will attempt to compile the newly created rule base, and will repeat the edit-compile cycle if errors are indicated by the compiler.

It is important to note that the only way that *AGASSISTANT* currently can acquire a hierarchically structured rule base is by directly creating it. This is because the current learning program used in the system is not capable of constructing a knowledge base with intermediate layers of knowledge. One can work around this, by learning a set of subconcepts, and then learning higher-order concepts or directly entering such concepts. For example, one could first learn a set of rules describing diseases that afflict a given species. One could then do another experiment to learn the best treatment for the plant with one of the variables in this experiment being the disease, if any, of the plant. One could then concatenate the two resulting rule sets within the editor to produce a structured knowledge base. This method can be used to produce a rule group of arbitrary depth, although it is likely that any complex domain will consist of a mixture of expert and induced rules.

Rule Learning

The underlying algorithm for all the learning facilities within AGASSISTANT's Inference Engine module is the NEWGEM program (Reinke 1984). At the heart of NEWGEM is the *Aq* quasioptimimal covering procedure. As *Aq* is described in detail in many previous papers, (see e.g. Michalski 1973) we will not go into great detail about it here. Suffice it to say that *Aq* works by attempting to find a rule which covers all of the positive events and none of the negative events, positive events being those belonging to the decision class under consideration, and negative events being all others. It does this by selecting a seed event within the set of positive events and extending it against successive negative events until it covers none of the negative events. Extending a partial cover against a negative event simply means specializing it so that it no longer covers the negative event if indeed it did to begin with. This process is continued until a cover or disjunction of covers is produced for all of the original positive events.

Learning Rules from Examples. Learning from examples is one of the most well-explored areas in machine learning (Dietterich and Michalski 1983; Michalski 1986). In this form of learning, a teacher provides characteristic examples and their respective decision classes to the learner. The task is then to create a set of rules which classify the given events. While simple in principle, this method of building descriptions of concepts is often powerful in practice. For example, in a now famous case it was shown that inductively derived rules for soybean disease diagnosis outperformed expert given rules (Michalski and Chilausky 1980a).

AGASSISTANT combines variable definitions as found in the variable table, events as found in the data table, and parameters as set in the parameters table to form the file SPECIAL.GEM which is then passed to the NEWGEM learning program. The parameters for the program determine various aspects of the rule creation process, including the breadth of the beam search used by the *Aq* algorithm, the lexicographic functional which is used in sorting candidate hypotheses, and the extent to which produced rules will be trimmed. (See Reinke (1984) for more information on the meaning of the NEWGEM parameters.) The NEWGEM module takes the input from SPECIAL.GEM and sends its output to the file LAST.GEM, which is then copied to the file {SYSTEM NAME}.RLE if the user decides to save the produced rules.

Each selector in a produced rule is associated with a weight. These weights are calculated by the following formula, and then normalized so that that the weights of a given complex sum to 100.

$$(4) \qquad weight = \frac{pe}{pe + ne}$$

where *pe* is the number of positive events covered, and *ne* is the number of negative events covered by the selector. Thus, the weight produced represents the probability that the given decision class is indicated given that the selector is satisfied.

Improving Rules with Examples. AGASSISTANT is capable of improving its knowledge as new examples are presented to it. This method is known as incremental learning with perfect memory, and the algorithm for performing this task is presented in detail in Reinke (1984). Summarizing his description, we find the method to be a straightforward extension of the *Aq* algorithm. First, the existing cover for a class of events is specialized to take into account new negative examples. This new cover is then used as the original seed event for *Aq*.

As can be seen in Figure 5.9, the input for rule improvement is constructed from the existing rule base, the table of examples (which includes the newly entered examples), and the variable table. This file is then sent to the NEWGEM learning program which detects the presence of input hypotheses and therefore runs *Aq* incrementally. As Reinke mentions, one must be careful when using this learning mode, since it is likely that the complexity of the output rules will increase, although the amount of this increase depends on the nature of the new examples. Clearly, if the new examples are for the most part in existing decision classes, the rules will not change that drastically assuming that the original induction was performed on a statistically large enough sample of events. If the new events fall into new classes, or if the original events came from a small subsection of the true problem space, then the rules will exhibit a proportional increase in complexity. The chief advantage of the incremental learning method presented here is the speed increase of the induction process. Incremental learning was only partially implemented in AGASSISTANT.

Rule Optimization. The system allows a user to optimize selected rules according to certain criteria. One form of this is a conversion of characteristic rules to discriminant rules. Michalski defines characteristic rules as those that specify common properties of the members of a class, and discriminate rules as those which have only enough information to distinguish one class from another or another set of classes (Michalski 1983). Here characteristic rules will refer to any that are not discriminant. For example, expert-created rules typically fall somewhere between the discriminant and characteristic categorizations. Indeed, one common use for this facility is to compress rules provided by an expert to their discriminant versions.

The method for rule optimization takes advantage of facilities already provided by the NEWGEM and submitted as input hypotheses with no corresponding input examples. If the rule type parameter is set to produce discriminant rules, these rules will be generalized to discriminant form.

Figure 5.10 shows the results of converting three characteristic rules to discriminant form. Notice that the optimized rules contain only the information necessary to distinguish between the three actions, in this case, the values of the variables shape and texture. This method of rule optimization only makes sense in the context of an expert system if one is confident that the values of the discriminatory variables will be known by the user of the expert system. If this is not the case, the system will perform better if the rules are left in their characteristic form.

FIGURE 5.10 Rules Before and After Optimization

Action is one if:	Action is two if:	Action is three if:
Before Optimization		
1. Color is red,	1. Color is red,	1. Color is red,
2. Size is small,	2. Size is small,	2. Size is small,
3. Shape is round,	3. Shape is square,	3. Shape is square,
4. Texture is smooth.	4. Texture is rough.	4. Texture is smooth.
After Optimization		
1. Shape is round,	1. Shape is square,	1. Shape is square,
2. Texture is smooth.	2. Texture is rough.	2. Texture is smooth.

Rule Compilation

Once acquired, rules must be compiled if they are to be used in an advisory capacity. The compiler has two chores. One is to check the syntax of the rules and provide the user with the appropriate error messages if the rules do not parse. If the rules parse successfully, the compiler will then create a file suitable for the Advisor module to ask questions and give advice. The exact nature of this file was detailed earlier and will not be repeated here.

Figure 5.11 below contains the complete grammar for rules in the system. Summarizing this figure, a rule consists of a condition part and an action part. A condition consists of the disjunction of a set of complexes, which in turn consist of the conjunction of selectors. A selector consists of a variable, a relation, and either a value, a disjunction of values, or a range of values, plus an optional weight.

The rules are parsed in a straightforward way, that is, the parser is structured as a finite-state machine, with each of the elements of the machine optionally being another sub-machine.

FIGURE 5.11 Grammar for Rules

< rule >	: =	< action > < condition >
< action >	: =	< variable > < relation > < value > "if:"
< condition >	: =	< complex > "OR" < condition >
	:	< complex >
< complex >	: =	< selector > < complex >
	:	< selector >
< selector >	: =	< variable > < relation > < value-list > < terminator >
	:	< a-expression > < a-relation > < a-expression > < terminator >
< variable >	: =	string of letters
< relation >	: =	string of letters
< value-list >	: =	< value > "or" < value-list >
	:	< value > "to" < value >
	:	string of letters
< terminator >	: =	< termchar >
	: =	< termchar > < weight >
< termchar >	: =	" "
	:	","
< weight >	: =	natural number
< r-number >	: =	real number
< a-relation >	: =	" <," " < =," " =," " >," " > ="

An Exemplary System: WEEDER

To illustrate the use of AGASSISTANT, we describe its use to develop an advisory system WEEDER. In the design of an effective weed-control program for managing turf it is first necessary correctly to identify the species of weed(s) present and to determine the extent of their population. Morse (1971) outlines five basic identification methods for determining unknown plant species: (i) Expert determination, which is generally regarded as the most reliable of all identification techniques. This method merely transfers the responsibility of identification to an appropriate expert. This service can be slow and costly, and is often limited by the availability of an expert. (ii) Immediate recognition, approaches expert determination and accuracy. This is the ability of an individual to recognize an unknown weed by past examples of identification. For some taxonomic groups and immature plants, however, this method of identification is very difficult and in all cases requires extensive past experience. (iii) A comparison of an unknown specimen with identified species or illustrations. It offers a rapid, simple diagnosis and is often useful for many weeds commonly found in native populations. (iv) An identification key which is based on the development of appropriate descriptive phrases of morphological or biochemical characteristics. Identification keys generally take the form of groupings of similar characters from which the user must select the character which best matches that present on the unknown sample. The selection of this character then leads to the next set of identifying characteristics. This process is followed until enough characteristics have been identified to suggest the identification of the specimen. (v) The last identification technique is a diagnostic table or polyclave. Diagnostic tables are a matrix of rows of species and columns of identifying characteristics. Users of a diagnostic table can identify the listed characteristics in any order they wish.

Morse (1971) lists two major faults of identification keys: (i) They require a user to utilize certain characteristics whether or not they are convenient or can be identified; (ii) They implicitly rely on rigid descriptions of specimens. Occasional variation in a population can cause gross misidentification.

The use of expert systems techniques offers a new, unique method for assisting with species identification. The relative merits of an expert or advisory systems is the ability to select answers or queries about characters that are available on the unknown specimen. They can operate on various levels of uncertainty providing a more efficient mechanism for identification, particularly for immature plants which are even difficult for experts to identify.

Development of WEEDER

In order to prepare the knowledge for use with AGASSISTANT, a matrix was developed including each potential grass weed. Eleven identifying characteristics, both vegetative and floral were determined for each weed. The information was obtained from many sources: textbooks, weed identification manuals, botanical manuals, and the authors' experience. Variable names, types, and sets of values is shown in Figure 5.12. The characteristics selected were those thought to be most easily recognized in the field without supportive equipment. (Shurtleff et al. 1987) Figure 5.13. presents a typical rule. This rule consists of 10 selectors. Each selector is associated with a weight or confidence level (CL) that indicates the relevant importance of the condition in making the decision. Notice that these weights need not add up to 100; they are normalized by the system. The weights give a rough estimate of the importance of each of the conditions in discriminating between the rules. These weights were then refined by the domain expert (T. Fermanian).

Rules for WEEDER were developed utilizing the NEWGEM module of *AGASSISTANT*. Separate rule sets were formed, first by inducing a set of characteristic rules, *m* and then by inducing a set of discriminant rules. The most appropriate rules from both sets were then modified, utilizing expert experience and written to a single rule set used in the initial evaluation.

Grass weed identification in turf is generally only available through the use of vegetative characteristics. This is due to the frequent mowing of the turf which often removes any floral portions of the plant. WEEDER allows the user to select either vegetative or a combination of floral and vegative characteristics at the beginning of each session. This is done through a "does-not-apply" question which is always asked first in the consultation.

"Does not apply" questions were established in order to provide a meaningful subset of variables for WEEDER to act on. The question "Are seedheads or flowers present?" to which the user responds "yes," "no," or "don't know" begins each session. If a "don't know" answer is given, then all identifying characteristics are asked. If "yes" is answered, then eleven of the possible characteristics are presented to the user. If "no" is answered, which is

FIGURE 5.12 Variable Names, Types, and Values for WEEDER

VARIABLE	Vernation	Auricle	Ligule	Sheath	Collar	Blade-width
TYPE	nominal	nominal	nominal	nominal	nominal	linear
1	folded	absent	ciliate	compressed	narrow	fine
2	rolled	short	round	round	divided	medium
3		claw-like	truncate	closed	broad	coarse
4			acute			
5			toothed			
6			acuminate			
7			none			
8						

	Habit	Glumes	Disart	Awns	Florets	Flower	Nerves
	nominal	nominal	nominal	nominal	integer	nominal	integer
	bunch	shorter	above	absent	1	panicle	1
	rhizome	longer	below	present	3	raceme	3
	rhiz-stolon			bifid	5	spike	5
	stolon						

FIGURE 5.13 A Rule for Identifying Stinkgrass

Weed is Stinkgrass if:	CL
1. Florets are 10 to 12,	85
2. Flower is panicle,	70
3. Collar is narrow,	60
4. Blade-width is medium,	55
5. Habit is bunch,	50
6. Sheath is compressed,	50
7. Vernation is rolled,	35
8. Glumes are shorter,	30
9. Florets is 1,	30
10 Disart is below.	25

FIGURE 5.14 Does Not Apply Conditions in WEEDER

If seedheads are present then
    Auricle, and Blade-width do not apply.
If seedheads are not present then
    Florets, Flower, Awns, Disart, and Glumes do not apply.

the usual situation for turf, then seven characteristics are presented—only those pertaining to vegetative portions of the plant. Paraphrased versions of these do not apply conditions are shown in Figure 5.14.

WEEDER Evaluation

In order to evaluate the absolute efficiency of WEEDER in drawing the correct conclusion, it was necessary to develop a program to determine the minimum number of variables required to identify each species. This program determined which groups of variables would provide a CL of threshold or greater value for a chosen rule. This program was run external to AGASSISTANT and was used to determine the maximum set of variable combinations for each rule in WEEDER.

Gower and Barrett (1971) state that the most efficient determination of an unknown species using an identification key is to use identifying variables which divide potential species into equal binary groups. They therefore suggested an equation to represent the minimum theoretical number of decisions necessary when using a dichotomous identification key.

Minimum number of decisions $= \log_2 n$

Where n represents the total number of species considered. If a dichotomous key was constructed to identify the 37 grasses of WEEDER, a minimum of five variable decisions would have to be made for a positive identification. This minimal number of decisions would therefore provide the most efficient use of the key. This would require each decision to equally divide the species which is not possible for a key using the chosen variables for the 37 species in WEEDER. In practice most identification keys do not provide this optimum efficiency (Pankhurst 1978).

Only the subset of variables describing vegetative characters was used in the evaluation of WEEDER. For the 37 grasses, there was a maximum of 16 sets of variables providing for the correct identification of any one grass. For four species only a single set of variables provided its identification. There were a total of 145 different sets of variables which represented correct identification. There were a total of 145 different sets of variables which represented correct identifications of any species with a mean of four sets for each species. The average number of variables necessary to correctly identify each grass species and its accompanying mean CL is listed in Figure 5.15. An identification was made when the CL was 85 or greater. The mean CL for all identifications was 92.

With a mean number of five variables required for each identification, WEEDER's efficiency was similar to the theoretical maximum efficiency for a dichotomous key to identify the same species. A dichotomous key with this level of efficiency has not been constructed using the same set of variables and values for the species in WEEDER. The maximum number of variables required for a correct identification of any species was seven. Over one-half (59%) of the identifications required five or fewer variables. For most species (96%), the recognition of a maximum of six variables was required for its identification. Since dichotomous keys generally do not perform at the theoretical maximum efficiency for identifying species, WEEDER shows excellent potential as a grass identification tool.

FIGURE 5.15 Mean Number of Variables and Average CL Required to Identify Any of 37 Grass Species Using WEEDER.[†]

	No. of variables[‡]	CL
Mean	5 [§]	92
Minimum	4	88
Maximum	6	99
SE[¶]	0.1	0.5
CV (%)[#]	11	3

[†] For each unknown specimen an identification was considered correct when the CL was ≥ 85.
[‡] Mean number of variables necessary for the identification of each species.
[§] Mean of 37 species means.
[¶] Standard error of mean.
[#] Coefficient of variation.

The WEEDER advisory system provided an excellent exemplary system and helped to test some of the proposed capabilities of AGASSISTANT. Specifically, WEEDER was able to provide a means for the satisfactory identification of 37 grass species commonly found in turf through a minimal number of decisions. In most cases, multiple sets of variables could be used to identify a single species.

Validation of WEEDER

In order to measure the relative efficiency of WEEDER in identifying unknown grasses, a study was conducted in which individuals were asked to identify four unknown grasses. Four grasses were selected randomly from a set of fifteen grass species commonly found in central Illinois. The four species selected were: creeping bentgrass (*Agrostis palustris* L.), perennial ryegrass (*Lolium perrenae* L.), zoysiagrass (*Zoysia japonica* L.) and large crabgrass (*Digitaria sanguinalis* (L.) Scop.).

Forty-one volunteers were assigned to one of two groups. If they had previous experience in plant diagnosis or formal training in plant science, they were separated from those volunteers who had no biological or plant science training or experience. Each individual randomly selected two of the four unknown weeds for identification using WEEDER, the other two weeds were identified using a diagnostic key, a commonly used tool (Shurtleff et al. 1987).

Along with the four unknown grass samples, each participant was supplied with a low-power dissecting microscope, appropriate probes and dissecting equipment, and a book with representative diagrams of all the potential configurations of morphological characters. Each individual was allowed up to 30 minutes per weed for identification. Fifteen minutes was reserved for a demonstration of each character and an explanation of how it could be identified. For the plants identified through the diagnostic key, each participant only supplied their first and possibly a second choice, as suggested by the key. Grasses identified with WEEDER, however, offered participants the ability to indicate the configuration chosen for each plant character. A frequency analysis of the correctly identified grasses was then conducted to determine their fit to the χ^2 distribution.

WEEDER has the ability to rank all the grasses in its knowledge base from the species most likely to represent the unknown grass to the one least likely. Figure 5.16 presents the mean percentage of identified grasses across all species using either identification tool (WEEDER or the identification key). The identification key, a tool commonly used by the participants in the study with plant science training, showed the highest average rate of success (20%) for identifying a species in the initial evaluation. The mean success rate of participants with plant science training in identifying any species using

FIGURE 5.16 Mean Percentage of All Correctly Identified Grass Species Using Either WEEDER or an Identification Key by Participants with Either Plant Science Training or Without Plant Science Training.

Selected Frequency Group	Mean correctly identified species	
	Initial rules	Modified rules
Identification tool		
WEEDER[†]	11	50
Identification key	20	20 [‡]
χ^2	2.3	16.8
	NS	**
Participant group		
Plant science training [§]	19	39
No plant science training	13	32
χ^2	2.7	1.9
	NS	NS

*,** Significant at the 0.05 and 0.01 levels, respectively. NS = not significant at the 0.05 level.

[†] For both participant groups.

[‡] Since the modification of an identification key is not practical the same values were used for the "Modified rules" evaluation.

[§] For both identification tools.

WEEDER was 15% and 7% for participants without plant science training using WEEDER (not shown in Figure 5.16.) After this initial experiment it would appear that the test group was not successful using WEEDER to identify the selected species. A closer evaluation of the answers selected by those using WEEDER showed a consistent problem in correctly determining the value of a few morphological characters. The natural variation in the growth of the selected species and their juvenile state made the identification of fine characters, such as the ligule, very difficult.

Rules for identifying the four grass species examined were modified (Figure 5.17) through the incremental learning facility in AGASSISTANT. The test groups answers were used as examples to improve the original rules. The results of these changes showed a very large gain in the percentage of correctly identified grasses. On the average for both groups, the percentage of correctly identified grasses rose from 11 to 50% when using WEEDER, as compared to the 20% mean for grasses correctly identified with the identification key (Figure 5.16.)

While no significant indication of dependence on either the identification tool or participant group was shown (Figure 5.16) using the initial rules, a very significant dependence ($P < .01$) on the identification tool used after rule modification indicates the potential advantage of WEEDER over the identification key for all participants. No significant indication of dependence on

Figure 5.17 Rules Representing Four Grass Species Used in the Evaluation of WEEDER Before and After Their Modification

Initial		Modified[†]	
Weed is Bentgrass if:	cl[‡]	**Weed is Bentgrass if:**	cl
1. Ligule is round,	65	1. Ligule is round or toothed,[§]	65
2. Sheath is round,	65	2. Sheath is round,	65
3. Glumes are longer,	65	3. Glumes are longer,	65
4. Habit is stolon,	60	4. Habit is stolon,	**40**
5. Disarticu is above,	55	5. Disarticu is above,	55
6. Collar is narrow,	50	6. Collar is narrow,	**70**
7. Florets is 1,	45	7. Florets is 1,	45
8. Flower is panicle,	45	8. Flower is panicle,	45
9. Blade-width is fine,	35	9. Blade-width is fine,	**70**
10. Vernation is rolled.	30	10. Vernation is rolled.	**70**
Weed is Per-ryegrass if:	cl	**Weed is Per-ryegrass if:**	cl
1. Ligule is round,	85	1. Ligule is round or **truncate**,	**25**
2. Auricle is short,	80	2. Auricle is short,	80
3. Florets is 6 to 10,	80	3. Florets is 6 to 10,	80
4. Flower is spike,	75	4. Flower is spike,	75
5. Vernation is folded,	50	5. Vernation is folded,	50
6. Habit is bunch,	40	6. Habit is bunch,	40
7. Collar is broad or divided,	35	7. Collar is broad or divided,	**75**
8. Sheath is compressed,	30	8. Sheath is compressed,	**70**
9. Blade-width is fine to medium,	30	9. Blade-width is fine to medium,	**70**
10. Disarticu is above,	35	10. Disarticu is above,	35
11. Glumes are shorter.	25	11. Glumes are shorter.	25
Weed is Zoysiagrass if:	cl	**Weed is Zoysiagrass if:**	cl
1. Habit is rhiz-stolon,	80	1. Habit is rhiz-stolon **or rhizome**,	80
2. Glumes are longer,	80	2. Glumes are longer,	80
3. Awns are present,	75	3. Awns are present,	75
4. Flower is spike,	70	4. Flower is spike,	70
5. Sheath is round,	70	5. Sheath is round,	70
6. Ligule is ciliate,	60	6. Ligule is ciliate,	60
7. Florets is 1,	55	7. Florets is 1,	55
8. Blade-width is medium,	50	8. Blade-width is **fine to** medium,	50
9. Collar is broad,	50	9. Collar is broad,	**70**
10. Disarticu is below,	45	10. Disarticu is below,	45
11. Vernation is rolled.	35	11. Vernation is rolled.	**75**
Weed is Lg-Crabgrass if:	cl	**Weed is Lg-Crabgrass if:**	cl
1. Ligule is toothed or acute,	65	1. Ligule is toothed or acute,	65
2. Blade-width is course,	60	2. Blade-width is **medium**,	60
3. Flower is spike,	60	3. Flower is spike,	60
4. Sheath is compressed,	50	4. Sheath is compressed,	50
5. Habit is bunch,	40	5. Habit is bunch,	**60**
6. Disarticu is below,	40	6. Disarticu is below,	40
7. Collar is broad,	35	7. Collar is broad **or divided**,	35
8. Florets is 1,	35	8. Florets is 1,	35
9. Vernation is rolled,	35	9. Vernation is rolled,	**75**
10. Glumes are shorter.	20	10. Glumes are shorter.	20

[†] Rules were modified after initial frequency analyses of chosen values

[‡] Confidence level assigned by system developers.

[§] Portions of the rules which were modified appear in **bold**.

participant group was found for correctly identifying a grass species after the rules were modified.

An analysis of the frequency of the selected values for each variable by either group of participants using either identification tool showed no significant dependency on individual values for any variable. In several cases, such as the toothed value of the ligule variable for bentgrass (selected for 53% of the bentgrass specimens) and the fine value of the blade width variable for zoysiagrass (selected 94%), the identified variable value was quite different than the one provided in the original rule. In addition, many of the variables which had low CL's in the original rules were most readily identified by the participants. For example, the original CL for the vernation-rolled zoysiagrass were 35 but was selected 82% of the identifications.

One of the most prominent findings of this investigation was the relatively poor performance in the identification of unknown grasses by individuals regardless of their training. Because the mean correct identification of any species by any participant group using WEEDER was less than 60%, it was not known if an expert level performance was achieved. In a subsequent study (Fermanian et al. 1989), considered over all characters, trained participants selected the correct value 59% of the time, whereas the untrained participants did so 53% of the time. No significant association was observed between participant groups and their selection ability for ligule size, sheath, blade width, collar, and pubescence when all species were considered jointly. Various programs have been developed for the identification of plant species by matching user selected values with similarity coefficients (Pankhurst 1975; Ross 1975). While these systems have generally reported similarity values of 60 to 90% ($\pm$ 5%) the success rate of identifying unknown species with the systems was not reported.

When using the identification key, performance was generally better from the group with plant science training, however, the frequency analysis did not indicate a significant dependance on either participant group. This difference in performance, however, was not found when the same group used WEEDER, which generally benefited either group equally. It is important to note that a significant gain in the ability of all participants to correctly identify a grass specimen was found with WEEDER over the diagnostic key, after rules were modified to maximize the support of constently chosen correct values of variables to identify the specimens examined. While the modification of a rule generally provided for the identification of specimens which were previously not identified, it also removed some specimen identifications from the group initially considered correct. Further testing of WEEDER is required to determine if the modified rules are consistent with additional grass sample and user populations.

This study brings out one important aspect to the use of expert or advisory sytems. While the use of knowledge is central to all advisory systems, the skills associated with recognizing the value of prompted variables is paramount in plant species identification. These recognition skills were probably lacking in the test population. It is necessary, therefore, to develop techniques to enhance recognition skills to further increase the effectiveness of WEEDER (Michalski 1986).

While WEEDER provided an initial test of AGASSISTANT's inferencing capabilities, other portions of the program remain untested (learning module, rule optimization, etc.). Additional efforts currently are being developed to test these functions.

Conclusions

An expert system builder that is capable of learning and improving its knowledge has been presented. Thus it has been demonstrated that sophisticated knowledge acquisition facilities are suitable for creating expert systems in the microcomputer environment. This should be of great use in disseminating this technology to the typical agricultural user who does not have access to large computers. However, A number of improvements and extensions to AGASSISTANT are possible.

- The system could incorporate new and more powerful learning subsystems. One such method is learning by analogy, in which the program acquires knowledge by comparing to similar cases it has seen in the past.

- The ideal system should be able to adjust its knowledge during the advisory session. That is, if it is told that it made an incorrect decision, it should be able to update its knowledge in light of this information.

- The current system uses a very simplistic method for combining evidence. The creator of the system should be able to state the importance of groups of conditions in addition to weighting individual conditions.

- The system could benefit from automated methods for generated explanation and other text during the advisory session.

- The learning module should be able to incorporate background knowledge. In addition, it should be able to suggest a hiearchical structure whereby input events are connected to decision classes through intermediate nodes.

- The system could be expanded to include the programs CLUSTER (Stepp, 1983), for clustering examples into categories, and ATEST (Michalski 1985), for testing the consistency and completeness of rules.

- Finally, the system could be integrated with a video system. This would enable the system to display plants and other items during the question answering phase of the advisory session

This list can help to guide further research to make AGASSISTANT a still more powerful and useful tool for agricultural decision making.

References

Dietterich, T., and R.S. Michalski. 1983. A Comparative Review of Selected Methods for Learning from Examples. In *Machine Learning: An Artificial Intelligence Approach*, ed. R.S. Michalski, J. Carbonell and T. Mitchell. Palo Alto: TIOGA Publishing Company: 41–81. .

Fermanian, T.W., M. Barkworth, and H. Liu. 1989. Ability of Trained and Untrained Individuals to Identify Morphological Characters of Immature Grasses. *Agron. J.* 81:918–222.

Gower, J.C., and J.A. Barrett. 1971. Selecting tests in diagnostic keys with unknown responses. *Nature.* 232:491–93.

Michalski, R.S. 1973. AQVAL/1 Computer implementation of a variable valued logic system VL1 and examples of its application to pattern recognition. *Proceedings of the First International Joint Conference on Pattern Recognition*, Washington, D.C.

_____. 1975. Variable valued logic and its application to pattern recognition and machine learning. In *Computer Science and Multiple Valued Logic: Theory and Applications*, ed. D.C. Rine. North Holland.

_____. 1983. A theory and methodology of inductive learning. In *Machine Learning: An Artificial Intelligence Approach*, ed. R.S. Michalski, J. Carbonell and T. Mitchell. Palo Alto: TIOGA Publishing Company.

_____. July, 1985. Knowledge repair mechanisms: Evolution vs revolution, ISG 8514, UIUCDCSF85946, Urbana, IL: Department of Computer Science, University of Illinois (and) June 1985. *Proceedings of the Third International Machine Learning Workshop.* Skytop: Rutgers University.

_____. 1986. Understanding the nature of learning. In *Machine Learning: An Artificial Intelligence Approach*, Vol. 2, ed. R.S. Michalski, J.G. Carbonell, and T.M. Mitchell. Los Altos, CA: Morgan-Kaufmann.

_____ and A.B. Baskin. August 8–12, 1983. Integrating multiple knowledge representations and learning capabilities in an expert system: The ADVISE system. *Proceedings of the 8th IJCAI.* Karlsruhe, West Germany.

_____ and J.B. Larson. 1983. *Incremental Generation of VL1 Hypothesis: the underlying methodology and the description of program AQ11.* ISG 835, UIUCDCSF83905, Urbana IL: Department of Computer Science, University of Illinois.

_____ and R.L. Chilausky. 1980a. Knowledge acquisition by encoding expert rules versus computer induction from examples: A case study involving soybean pathology. *International Journal for Man-Machine Studies.* 12:63–87.

—— and R.L. Chilausky. 1980b. Learning by being told and learning from examples: An experimental comparison of the two methods of knowledge acquisition in the context of developing an expert system for soybean disease diagnosis. *International Journal of Policy Analysis and Information Systems.* 4(2):125–160.

——, J.H. Davis, V.S. Bisht, and J.B. Sinclair. 12–14 July 1982. PLANT/ds: An expert consulting system for the diagnosis of soybean diseases. *Proceedings of the 1982 European Conference on Artificial Intelligence.* Orsay, France. 133–138.

Morse, L.E. 1971. Specimen identification and key construction with time-sharing computers. *Taxon.* 20(1):269–82.

Pankhurst, R.J. 1975. Identification by matching. In Biological Identificaton with Computers. ed. R.J. Pankhurst. London & New York: Academic Press.

——, 1978. *Biological identification.* Baltimore, MD: University Park Press.

Reinke, R.E. July, 1984. Knowledge acquisition and refinement tools for the ADVISE MetaExpert System. M.S. Thesis ISG 844, UIUCDCSF84921, Urbana, IL: Department of Computer Science, University of Illinois.

Ross, G.J.S. 1975. Rapid techniques for automatic identification. In *Biological identification with computers.* ed. R.J. Pankhurst. London & New York: Academic Press.

Shurtleff, M.C., T.W. Fermanian, and R. Randell. *Controlling Turfgrass Pests.* Englewood Cliff, NJ: Prentice Hall.

Stepp, R. November, 1983. A Description and User's Guide for CLUSTER/2, A Program for Conjunctive Conceptual Clustering, Report No. UIUCDCSR831084, Urbana, IL: Department of Computer Science, University of Illinois.

6

Knowledge Acquisition: A Case History of An Insect Control Expert System

Pierce Jones, James W. Jones,
Howard Beck, and P.A. Everett

Introduction

During the last few years a new technology with tremendous conceptual potential has burst upon the engineering scene. Expert Systems, a subset of the field of Artificial Intelligence, has been heralded as a near-term practical problem-solving tool that can operate in areas forbidden to traditional algorithmic programming. While it is true that great progress has been made in the hardware and software needed to support expert system development, it is well-recognized among practitioners that the principle and non-trival bottleneck to the rapid development of expert systems is knowledge acquisition (Buchanan et al. 1983; Duda and Shortliffe 1983). Nevertheless, there has not been any systematic research on the question of how to elicit an expert's knowledge and inference strategies (Hoffman 1987). As a result the technology has been criticized, among other things, for requiring far too much development time to have any economically practical value (Dreyfus and Dreyfus 1986). Even so, a wide range of projects have been reported in the literature with titles that suggest significant practical progress (Jones and Haldeman 1986; Jones et al. 1985; Hoshi and Kozai 1984; Nelson 1982).

In the literature most of the current work presented on expert systems describes prototypic systems with very limited field testing, if any, and the

The authors would like to thank F. W. Johnson and R. K. Sprenkel for their time and on-going efforts serving as domain experts for the SOYBUG project and also J. Kerns for her steady help in the preparation of this manuscript. Mention of company names or commercial products does not imply recommendation or endorsement over others not mentioned.

general focus in professional meetings of agricultural scientists on the subject of expert systems has been on understanding the jargon, watching "shells" develop into expert system development environments, and discussing the relative merits of microcomputer-based systems vs. LISP machines. These are important initial concerns that deserve technical consideration, but most of this discussion of broad concepts, software, and hardware has tended to ignore the underlying problem of knowledge engineering: the effective extraction of knowledge from experts.

Now, that a wide range of software and hardware to facilitate expert system development is readily available and the jargon is becoming demystified, it is time to give the knowledge acquisition process the attention that it deserves. Figure 6.1 schematically relates the people (domain expert, knowledge engineer, and user) to the processes/components that make up an expert system. The knowledge acquisition component in the schematic is enlarged to emphasize its preponderant role in expert system development. It is also distinct from the other processes because it is not a software component, it is a truly qualitative process.

With their first systems, many budding knowledge engineers avoid the more difficult aspects of knowledge acquisition by choosing problems for which they can serve as their own expert (Jones and Haldeman 1986).

FIGURE 6.1. A Schematic Representation of an Expert System Development Environment.

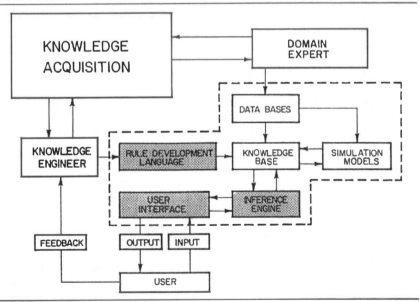

Note: The processes inside the dashed lines are software components; the shaded ones are generic shell software.

Although projects in which the knowledge engineer is his own expert provide useful on-the-job training with the software and hardware components of expert system development, it obscures the knowledge acquisition process. For this and other reasons it is generally agreed among the most experienced expert system builders that the roles of knowledge engineer and domain expert should be separated (Buchanan et al. 1983). Initially in such projects, the expert seldom knows much about expert systems and the knowledge engineer seldom has any expertise in the project's domain. Knowledge acquisition is the process of transferring knowledge from the domain expert to the knowledge engineer. This process can require a great deal of time depending on the degree to which the expert's knowledge is to be represented. For example, to obtain a 10% increase in performance, PUFF, an expert system for interpreting pulmonary test data required an increase from 100 to 400 rules (Aikens et al. 1983). Figure 6.2 is a graphical representation of the conceptual relationship between the time required in system development versus the degree to which the domain expert's knowledge is represented. As a system matures it can be expected to asymptotically approach the expert's performance. The time-scale depends on the complexity and extent of the particular problem. Waterman (1986) estimates that moderately difficult projects, such as PUFF, require about five person-years for development while very difficult projects such as MACSYMA (Martin and Fateman 1971) can require 30 person-years. This may be somewhat misleading because the recently available tools for expert system development may save steps. However, the qualitative knowledge acquisition process has not been cracked and will remain the major bottleneck to rapid development of expert systems for the foreseeable future.

The general purpose of this paper is to offer a candid review of a seemingly straight-forward expert system development effort. In particular, the paper's emphasis is on the acquisition of knowledge from experts outside the knowledge engineering/programming group.

Knowledge Acquisition

Conceptually, an expert system's knowledge base can be derived from a wide range of sources such as textbooks, case studies, simulation models, and anecdotes. However, the preponderance of knowledge in most well-developed expert systems is based on the informed opinion of domain experts. The knowledge engineer extracts and formulates an expert's heuristic model of a domain through "a prolonged series of intense, systematic interviews, usually extending over a period of many months" (Waterman 1986, 152). The experiences of early workers in this field have revealed several generally recognized difficulties that characterize this interview process.

The central difficulty that experts have is their inability to define the procedures or rules that they actually use in reaching their conclusions. They do not operate at a basic level, but typically see patterns or have hunches which they may attribute to intuition. This perception makes it difficult for the expert to articulate a step-by-step analysis of his procedure (Waterman and Jenkins 1979). In fact, it appears that the more expert an individual is, the more difficulty he has in defining his procedures. This phenomena has been called the paradox of expertise (Johnson 1983). As a sequel to the paradox of expertise, knowledge engineers have found that when experts attempt to explain their thought processes, they often provide rationalized lines of reasoning that have little in common with their actual decision-making process (Johnson 1983). In fact, an expert's working knowledge seems to be a composite of patterns gleaned from direct experiences in his domain rather than logically deduced applications of theory. Their knowledge seems to be compressed and refined into a compiled form. Waterman (1986) summarized the implications and difficulties caused by the paradox of expertise with three maxims concerning the knowledge acquisition process:

> The more competent domain experts become, the less able they are to describe the knowledge they use to solve problems!
>
> Don't be your own expert!
>
> Don't believe everything experts say!

These maxims strongly suggest that an expert would most likely have great difficulty extracting his own decision-making processes and would benefit from outside help in documenting his actual problem-solving procedures.

Given these general warnings, how should domain knowledge be most effectively extracted? Buchanan et al. identify generalized stages in the acquisition process, but state that it "is not understood well enough to outline a standard sequence of steps that will optimize the expert system building process (1983, 140)." Nevertheless, there are techniques that have gained some recognition. They are summarized by Waterman (1986) as follows:

1. Observe the expert solving real problems on the job.
2. Discuss the kinds of data, knowledge, and procedures required for particular types of problems.
3. Develop descriptions with the expert of prototypical problems associated with each category of answers in the domain.
4. Have the expert solve a series of realistic problems aloud, ask for the rationale behind each decision step.
5. Solve problems provided by the expert using rules developed from the interviews.
6. Have the expert review and criticize the rules and control structure.
7. Have outside experts solve problems already presented to the prototypic system and expert.

This list forms the knowledge acquisition framework that should direct the progression of interviews in the development of an expert system. Depicted graphically in Figure 6.2, as knowledge approaches expert knowledge, the extraction time requirement increases rapidly. This suggests the importance of setting realistic limits on the degree to which the system represents expert knowledge.

The balance of this paper is devoted to a description of the development of SOYBUG, an expert system for making spray recommendations for pests in soybeans. The emphasis of the paper is on the practical aspects of knowledge acquisition within the larger context of the expert system project.

Project Chronology and Discussion

Project Definition

The project was first conceived in March 1985 with the primary conceptual purpose demonstrating the potential for coupling heuristic expert systems and algorithmic simulation models. A secondary conceptual purpose was to demonstrate that inexpensive, commercially available shells for use on microcomputers could support relatively sophisticated expert systems.

INSIGHT2+ from Level Five Research was used for the project because of its capability to exit the expert system, run an external program written in a high level language, such as Turbo Pascal (Borland 1985), and return to the expert system. Based on this feature the project's primary software goal was to use heuristically defined values to parameterize a simulation model, to pass values generated by the simulation back to the expert system, and to make a decision.

FIGURE 6.2. A Graphic Representation of the Time Required for Knowledge Extraction.

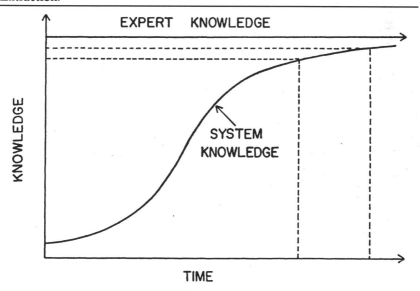

The general goal of the expert system was to make spray decisions in soybeans. Specific expert system goals were not initially defined. Because of previous work with a well developed simulation model of soybean/pest interactions, SICM (Wilkerson et al. 1983), and the presence of several entomologists with expertise on this topic, the project's domain was defined to be spray decisions for pests in soybeans with velvetbean caterpillar (VBC) being the primary pest of interest. Other pests were to be considered in combination with VBC. It is instructive that the software goals were the project's initial focus.

Initial Interviews with the Domain Expert

During June 1985, the first contact was made with F. A. Johnson who became the project's first domain expert. He was chosen because of his years of experience as an extension specialist in the problems of pest control in soybeans. He agreed to work on the project although he wasn't sure how he could help us. This attitude reflected a generally imprecise understanding of the various roles to be played that existed early in the project.

At the initial interview, he had no knowledge of expert systems, but was very familiar with the concept of interactive information retrieval. This may have contributed to some of the confusion which became apparent when the discussion shifted to a delineation of the domain. The domain required a very narrow definition for purposes of the expert system and because the simulation model (SICM) was well-validated only for VBC (Wilkerson et al. 1983). As the interview shifted to questions concerning the expert's procedures for making recommendations, it became apparent that he found it difficult to narrow his focus and not deal with broad possibilities. The most useful aspects of the first interview could be categorized as parameter listing, anecdotes, and telephone technique.

Following the initial interview extension bulletins (Johnson et al. 1984a; Johnson et al. 1984b) that made specific suggestions concerning pest control in soybeans were reviewed. Based on this information and on the first interview, scenarios were developed to determine when the experts might deviate from 'standard recommendations'. Also, the telephone format was developed; the expert was to respond to the scenario as if it were a telephone call from a farmer who wanted to know whether or not he should spray. In fact, consideration was given to the idea of actually conducting the interviews over the telephone.

The second interview was tape recorded. In many respects this interview seemed repetitive of the first. However, its format was somewhat different, a general purpose scenario was put forward in the telephone format. Although it became apparent that the scenario and its presentation were not sufficiently complex to actually engage the expert, it did provide some needed focus to the discussion. The interview revealed a parameter list that seemed very straight-forward, but the order and content of the list seemed to shift in subtle ways not readily defined at the time.

In both interviews the expert referred to extension bulletins and the opinion of other experts. He specifically requested that another pest specialist be brought into the process. As a result R. K. Sprenkel, an extension pest specialist located at a research station in the soybean growing region of Florida, was asked and agreed to participate as the project's second domain expert. Several more interviews with each of the experts took place during the summer, but diminished during the fall.

Algorithmic Modeling

In August 1985 the modeling effort began to take shape when the decision-making process was interpreted as simply comparing control costs all season with increased crop value, i.e. an economic basis. As a result, an economic simulation model that operated outside the expert system was developed, written in Turbo Pascal. As the modeling effort grew, interactions with the experts diminished, and the heuristic portion of the expert system began to disappear. Inputs to the economic model that were initially conceived to come from heuristic rules were obtained from databases developed from simulations using SICM. The primary remnant of a heuristic model was in rules for choosing specific pesticides. By November the project's effort was to round off the rough edges and write up the results for presentation at a professional meeting (Jones et al. 1985). This was an important turning point in the project because it required that the goals and methods of the project be thoroughly evaluated.

As reported by Jones et al. (1985) the project's original software goals were met by December, 1985. Heuristic and algorithmic models (and databases) were successfully coupled in a prototypic expert system. The system was developed with inexpensive shell software and was running on a microcomputer. The paper revealed the extent to which modeling dominated this phase of the project. In the section on knowledge acquisition three times as much space was devoted to explaining knowledge developed from simulation runs and the economic model as was devoted to discussing knowledge extracted from the domain experts. This disparity was tacitly recognized in one of the paper's conclusions, if an *operational* (expert) system is the goal, the proper approach would be to carefully identify the entire problem at the outset. Then, work closely with the experts to learn the existing decision making process, including the full set of possible decisions or recommendations and the information, rules, and logic used to reach a decision. The specialist may use models as a routine part of the process and these should be identified. Then, the system should be developed and tested using as closely as possible the process used by the expert. Adding new models or logic not currently used by the expert should be considered applied research.

Redefinition of Expert System Goals

In January, 1986 the project's focus shifted to the preparation of a second paper (Jones et al. 1986) presented at the International Conference on Computers in Agricultural Extension Programs which took place in February, 1986. This presentation emphasized the development of an operational expert system with the goal of being able to make recommendations comparable in quality to those of the specialist whose knowledge is represented. In this paper some of the key conceptual procedures in expert system development were outlined; from a schematic of an expert system, through a decision tree representation of knowledge, to an example of an IF/THEN rule. As co-authors, both of the domain experts critiqued the paper with the result that they developed a fuller appreciation of their role and potential significance in building the expert system. This coincided with the successful completion of the original software goals (in December) and a consensus of opinion on the knowledge engineering side that algorithmic models would be held in abeyance until reasonably complete hueristic models were developed. For these reasons SOYBUG gained a new momentum with a new more purely heuristic direction.

Second Interview Phase

Following the February presentation regular weekly interviews were initiated by one of the domain experts. Progress in defining the domain and focusing on the more pertinent parameters came quickly, leading to a conceptual model of the decision process. The spray decision process was subdivided into four major problem areas; 1) identification of the insect pests (Have the insects been correctly identified?), 2) quantification of insect population (Are population estimates accurate?), 3) delineation of yield damaging population thresholds (What population should trigger a spray decision?), and 4) choice of pesticide (For given circumstances which is the safest, most effective and most economical pesticide?). Of these four the pivotal point seemed to be the expert's changing perception of insect population spray thresholds in differing scenarios. So, the interviews began to focus on that issue.

As had been previously recognized the rule-of-thumb thresholds published in the extension bulletins were economically conservative. The philosophy behind extension threshold recommendations was that it is better to spray a little too often than to risk a major economic loss. This implied that given complete information for scenarios with insect populations near the extension threshold where the bulletin would recommend spraying the experts might recommend against spraying. This realization shaped the first clear and adequate definition of the project's goal for an operational expert system: When given a scenario, SOYBUG should always be consistent with the experts' recommendations to spray and should never recommend spraying if the extension bulletin does not. This places SOYBUG between the static broad rule-of-thumb extension recommendations and the subtle complete information recommendations of the experts who wrote the bulletin.

Development of Specific Observation Technique

With this goal in mind interviews became increasingly specific culminating in a listing of scenarios near the rule-of-thumb extension thresholds (see Figure 6.3). In March, 1986, the scenarios were given to each expert for their recommendations. Recommendations were also developed for the scenarios using SOYBUG and the extension bulletin. A comparison of the recommendations gave some quantitative evaluation of our progress, but more important were the experts' qualitative concerns about the information in the scenarios. In the first set of scenarios the VBC population was broken into three size categories; small, medium, and large. In discussing their recommendations the experts were very concerned about both ends of the size scale, very small or small, very large or large. According to the experts, this finer distinction was important because very small VBC were subject to predation and very large VBC were near pupation. So, both extremes were not likely to remain in the field and might be ignored in deciding whether the population would reach the spray recommendation threshold. Additional questions were raised about crop value and a general explicit evaluation of crop health.

In response to these concerns, in April 1986, the same scenarios with five size categories (very small and very large were added) and additional information on crop value were given to the experts and evaluated according to a revised version of SOYBUG, and the extension bulletin. As expected, spray recommendations from the experts and SOYBUG did shift while the bulletin's remained static. When one of the experts took the 'test' it was noticed that he drew a line between the small and medium categories. On questioning, it became clear that he did not give equal weight to the small, medium, and large categories, although equal weight was given in the bulletin's formula.

Refinement of Observation Technique

In April 1986 the knowledge engineering/programming role shifted from one person to another. Naturally, a number of changes resulted from the different perspectives introduced by a new knowledge engineer. Based on several orientation interviews with the experts and an examination of the first two validation tests, greater emphasis was placed on studying case histories and examples. It was decided that the validation tests needed to be expanded in several ways. Changes made included: expanding each scenario to include a fuller description of conditions and assumptions; picking scenarios to cover a broader range of conditions (not just borderline cases); letting the experts fully describe a recommendation rather than select recommendations from a multiple choice list; letting them ask additional questions, and list any assumptions made. These changes led to the development of a new test (Figure 6.4).

FIGURE 6.3 A Simple, Specific Set of Scenarios in Multiple Choice Test Format.

Health[a]	No. Insects/row foot[b] vsm/sm/med/lg/vlg					Plant Stage	spray	wait 4 days	wait 6 days
excellent	3.5	3.5	2	4	4	prebloom	____	____	____
good	3	3	2	.5	.5	postbloom	____	____	____
poor	1.5	1.5	2	.5	.5	postbloom	____	____	____
excellent	4.5	4.5	9	--	--	prebloom	____	____	____
good	4	4	6	1.5	1.5	prebloom	____	____	____
poor	4	4	8	.5	.5	prebloom	____	____	____
excellent	2.5	2.5	4	--	--	postbloom	____	____	____
good	1	1	7	.5	.5	prebloom	____	____	____
poor	5	5	-	1	1	postbloom	____	____	____
excellent	-	-	8	1	1	prebloom	____	____	____
good	1.5	1.5	7	-	-	prebloom	____	____	____
poor	3	3	6	1	1	prebloom	____	____	____
excellent	2.5	2.5	3	-	-	postbloom	____	____	____
good	1	1	1	1	1	postbloom	____	____	____
poor	-	-	1	1	1	postbloom	____	____	____
excellent	-	-	2	1.5	1.5	postbloom	____	____	____
good	5	5	6	1	1	prebloom	____	____	____
poor	2	2	1	2.5	2.5	prebloom	____	____	____
good	2	2	-	1	1	postbloom	____	____	____

Assume average crop value of $6.25–$7.00 per bushel.

NOTE: Explicit scenarios near the broad rule-of-thumb spray threshold were used to uncover subtle distinctions involving size distributions in the insect population, plant stage, and crop health.

[a]Health	Percent defoliation Prebloom	Postbloom
Excellent	< 10%	< 5%
Good	10–20%	5–10%
Poor	> 20%	>10%

[b]Insect Size in inches	
Very Small	< 1/4
Small	1/4–1/2
Medium	1/2–3/4
Large	3/4–1–1/2
Very Large	> 1–1/2

FIGURE 6.4 An Example of a Broader Qualitative Scenario.

SCENARIO 3

FIELD CONDITIONS

VBC is present (*vsm* = 6/row foot, *sm* = 3, *med* = 3, *lrg* = O, *vlrg* = O). No other insect pests are present. Previous week's scouting report showed VBC present with

(*vsm* = 2/row foot, *SM* = 4, *med*-O, *lrg*-O, *vlg*-O).

Crop has no other stress except for 10% defoliation at present. It is pre-bloom and the canopy is nearly closed on 36" row spacing. Price of beans is average ($6.25–$7.00). The grower is comfortable taking reasonable levels of risks.

IS THIS SCENARIO REALISTIC?

OTHER QUESTIONS TO CONSIDER:

What if the grower did not like to take any risks, or
What if the grower could afford to take a high deal of risk.
What do you think the % defoliation will be in 7 days if this population is not controlled?

WHAT OTHER INFORMATION WOULD YOU LIKE TO HAVE TO MAKE THIS DECISION, WHAT ASSUMPTIONS ARE YOU MAKING?

RECOMMENDATION?

REASONS FOR RECOMMENDATION?

NOTE: This scenario which is #3 in a series of 50 requests a short answer response and encourages the expert to elaborate on his thoughts. These scenarios can be followed with interviews for deeper probing.

Project Evaluation

After more than a year of involvement with this particular project we have developed some impressions of how such an effort should be organized. Although the specifics of every expert system development will be unique, there are certainly some underlying generalities. With this in mind, an evaluation of our project's organization and of the interactions between the knowledge engineering and domain expert sides may prove useful to similar efforts.

Operational Goals

Originally, the SOYBUG project was guided by conceptual software goals which were predicated on the belief that the ideal computerized decision program would combine expert heuristic knowledge, experimental data, simulation models, and any other relevant information to make the best recommendation. Indeed, this still would be the ideal decision support tool. However, from the perspective of expert system development our operational position is that construction of such hybrid systems must be approached with care. We have constrained ourselves to building a system based completely on our domain expert's heuristic knowledge because current expert practice does not include models directly. This is not to say that the expert's model cannot contain algorithms, only that the use of models must arise from the expert's actual problem-solving procedures.

It is difficult always to separate existing and accepted problem solving procedures from those that could improve the conclusions or recommendations. For example, the experts in this project do consider the economic advantages for a particular decision. Our initial efforts (Jones et al. 1985) used a simple economic model to describe that aspect of the decision in a much more explicit form than was practiced by the experts. This model, in fact, could lead to improvements in arriving at economically sound decisions for pest control in soybean. However, this methodology was a new concept to extension experts and was not immediately accepted as a substitute for their own heuristic approach. In retrospect, this seems quite predictable. One should not expect to improve a procedure and have an expert immediately endorse it as his own. A new procedure or concept requires testing and thus falls into a category of technology development rather than knowledge (or technology) representation. We learned this the hard way, but now view this as a positive lesson. After the current procedures are represented and our expert system meets operational goals, the knowledge acquisition phase of the project can lead directly into a technology development phase, i.e., a new research phrase for incrementally improving existing procedures. Thus, perhaps these procedures will help to develop a more meaningful link between extension and research activities.

Our project's current goal is to build an expert system that comes very close to mimicking the reasoning processes and recommendations of our domain experts. Specifically, SOYBUG should substantially outperform recommendations based on extension bulletin rules. The degree to which SOYBUG outperforms the bulletin can be measured by the frequency of agreement between the experts and SOYBUG for scenarios that lead the experts to deviate from bulletin recommendations. Among its other advantages, this goal makes project development more tractable by clearly defining a stopping point.

Knowledge Acquisition Technique

Interviews are the foundation of any effort to extract and formulate expert heuristic knowledge. There is no avoiding the fact that any actual expert system development effort will require many lengthy and detailed interviews. Furthermore, the quality of these discussions is not consistent with traditional engineering analysis techniques or indeed with academic rigor generally. This difference is expressed in the following quote from Waterman: "It is seldom effective to ask the expert directly about his or her rule or methods for solving a particular type of problem in the domain (1986, 153)." To a certain extent the knowledge engineer must learn to see the patterns that the domain expert sees and must avoid looking for a step by step analytical framework.

Several aspects of the initial discussions were particularly helpful in revealing the expert's 'reasoning' processes and in educating the knowledge engineers about the nature of the problem domain. In retrospect these features can be viewed as prototypic techniques to be developed for future applications.

1. Tape recordings should be made during interviews and should be reviewed between interviews. This will help to avoid redundancy which can become particularly tedious for the expert.
2. The expert should be asked directly for facts (parameters) that he believes are essential or helpful in reaching decisions in his general area of expertise. It is not necessary to become overly specific about the domain during early interviews; instead a general extensive parameter list should be developed.
3. The expert should be asked for any broad or commonly accepted rules-of-thumb which may exist. Within agriculture, such heuristic rules are often documented in extension bulletins. These should be thoroughly considered.
4. The expert should be encouraged to relate any favorite or typical or instructive anecdotes about his area of expertise. Such stories tend to organize and connect parameters into coherent patterns of cause and effect.
5. Experts who often work with clients by telephone should be asked to create typical telephone scenarios. They should describe who their

client might be, what they might already know about the client, what information they might receive, and so forth.

6. Based on the first five procedures, the knowledge engineer should always bring something to the interview. Initially, it could be a verbal scenario, a scripted telephone conversation, or procedural schematics. Later, decision trees that lead to specific recommendations should be presented to the expert for criticism.

These are techniques that can be useful in the earlier interviews. During this phase the narrowness of the domain should not be emphasized if it constrains the expert. Instead, the early discussions should be viewed as requisite broad training of the knowledge engineer and as necessary for developing *rapport* between the participants. In this regard the interviews should be held regularly with time enough between them to allow the knowledge engineer to review and prepare for each interview.

In later interviews the techniques can be directed more specifically at the project's narrowly defined operational goals. The timing of this switch in orientation depends on the development of a mutual appreciation between the expert and engineer of their respective roles in the project. This should coincide to some extent with the unveiling of an operational prototypic expert system.

7. As soon as a debugged prototypic version of the expert system is available with a reasonable user interface, it should be shown to the expert. The expert should be encouraged to comment aloud as he runs the system. This procedure should be repeated as versions evolve.

8. The expert should be presented with simple, specific, pivotal scenarios in which secondary parameters vary. This serves as a sensitivity analysis for identifying crucial interactions between variables (see Figure 6.3).

9. The expert should be provided with more robust scenarios that encourage open-ended comment as opposed to simple answers (see Figure 6.4). Repetition of scenarios at different times can also reveal the stability of recommendations.

These techniques should help develop more explicit rules for unusual or close call cases. In all three techniques the expert should be asked to explain their comments or recommendations.

Finally, there are several techniques currently being devised that are expected to be useful for calibration and validation of the system.

10. With the expert's consent, several of his consultations in the domain could be recorded and transcribed for protocol analysis (Waterman and Jenkins 1979). This may be a bit tricky since it is hard to predict when pertinent calls will occur.

11. Without the expert's direct knowledge, scenarios from 'clients' can be planted. Such conversations could provide the most accurate response possible to pivotal scenarios. Several cases of this type could be very useful for validation.

These latter techniques have the potential to approximate most closely direct on-site observation of an expert.

Items 1-11 above summarize the procedures that we currently believe to be most useful in extracting heuristic knowledge to incorporate in expert systems like SOYBUG. The overall goal of techniques is to optimize the knowledge acquisition process; the transfer of domain knowledge from the expert to the engineer. However, such techniques also affect favorably the attitude of the knowledge engineer, particularly neophytes. Having specific tasks to perform in the early interviews can help to define the engineer's 'role' while he or she is being educated in the project's domain.

Conclusions

The purpose of this paper has been to analyze the knowledge acquisition process as it evolved in our particular project. Foremost among our conclusions and explicit in our analysis is the recognition of the absolutely central role of knowledge acquisition in the development of actual expert systems. In reviewing this project several key considerations concerning this essential process were identified.

First, recognize that the time commitment required depends primarily on the knowledge acquisition process and that software and hardware tools will not appreciably shorten overall system development time. Second, define a realistic operational goal for the expert system as soon as possible and explicitly review it frequently. Use algorithmic models only as they are used by the expert for an initial operational system. Third, identify the roles of project participants in terms of the operational goal and organize the project accordingly. For an operational system assume that the domain expert is the final authority who must accept the system.

Lastly, facilitate explicit feedback among the participants. In particular, the knowledge engineer should always bring something to the expert based on previous interviews such as explicit scenarios, decision trees, or specific rules for the expert's comments. Attending to these lessons learned about the knowledge acquisition process will enhance the chances for developing successful expert systems applications.

References

Aikins, J.S., J.C. Kunz, and E.H. Shortliffe. PUFF. 1983. An expert system for interpretation of pulmonary function data. *Computers and Biomedical Research.* 16:199–208.

Beck, H., J.L. Stimac, F.A. Johnson. 1984. Florida Agricultural Information Retrieval System (Fairs): Design of a computerized consultation system for agricultural extension. Ext. Bul. #227. Gainesville, FL: Institute of Food and Agricultural Sciences, University of Florida.

Borland International. 1985. *Turbo Pascal, Version 3.0. Reference Manual.*

Buchanan, B.G., D. Barstow, R. Bechtel, J. Bennett, W. Clancey, C. Kulikowski, T. Mitchell and D. Waterman. 1983. Constructing an expert system. In *Building Expert Systems.* ed. F. Hayes-Roth, D.A. Waterman, D.B. Lenant. Reading, MA: Addison-Wesley Publishing Company.

Dreyfus, H.L. and S.E. Dreyfus. 1986. *Mind over machine.* New York: McMillan.

Duda, R.O. and E.H. Shortliffe. 1983. Expert systems research. *Science* 220:261–276.

Hoffman, R.R. Summer, 1987. The problem of extracting the knowledge of experts from the perspective of experimental psychology. *AI Magazine.* 53–67.

Hoshi, T. and T. Kozai. 1984. Knowledge-based and hierarchically distributed online control system for green house management. *Acta Horticulturae.* 148:301–308.

Johnson, F.A., D.C. Herzog, and R.K. Sprenkel. May, 1984a. Soybean insect control. *Plant Protection Pointer* (Ext. Entomology Report #58). Gainesville, FL: Institute of Food and Agricultural Sciences, University of Florida.

_____, R.K. Sprenkel, R. Kinlock, and D.H. Teem. 1984b. *Managing Pests on Field Crops: Soybeans. Operation: IPM.* Florida 4H Integrated Pest Management Series. Extension Publ. #4H-377.8. Gainesville, FL: Institute of Food and Agricultural Sciences, University of Florida.

Johnson, P.E. 1983. What kind of expert should a system be? *The Journal of Medicine and Philosophy.* 8:77–97.

Jones, J.W., P. Jones, and P.A. Everett. 1985. Applying agricultural models using expert systems concepts. ASAE Technical Paper #85-5517.

Jones, P. and J. Haldeman. 1986. Management of a crop research facility with a microcomputer-based expert system. *Tran. of the ASAE.* 29(1):235–242.

Jones, P., J.W. Jones, P.A. Everett, F.A. Johnson, and R.K. Sprenkel. 1986. Development of an expert system for making insect control decisions in soybeans. *Proceedings at the International Conf. on Comp. in Agricultural Extension Programs.* Gainesville, FL: University of Florida.

Level Five Research, Inc. 1984. *INSIGHT,* 4980 South A-1:A, Melbourne Beach, FL.

Martin, W.A. and R.J. Fateman. 1971. The MACSYMA system. *Proceedings of the Second Symposium on Symbolic and Algebraic Manipulation.* 59–75.

Nelson, W.R. 1982. REACTOR: An expert system for diagnosis and treatment of nuclear reactor accidents. *Proceedings of Am. Assoc. for Artificial Intelligence.* Pittsburgh, PA: Carnegie-Mellon University.

Waterman, D.A. and B. Jenkins. 1979. Heuristic modeling using rulebased computer systems. In *Terrorism: Threat, Reality, Response.* ed. R.H. Kupperman and D.M. Trent. Stanford, CA: Hoover Institution Press.

Waterman, D.A. 1986. *A Guide to Expert Systems.* Reading MA: Addison-Wesley Publishing Co.

Wilkerson, G., J. Mishoe, J. Jones, J. Stimac, W. Boggess, and D. Swaney. 1983. SICM: Soybean Integrated Crop Management Model: Model description and user's guide, version 4.2., Agricultural Engineering Department Research Report AGE-1. Gainesville, FL: Agr. Engr. Department, University of Florida.

.

7

Knowledge Engineering: Creating Expert Systems for Crop Production Management in Egypt

Ahmed A. Rafea,
Merrill E. Warkentin, and Stephen R. Ruth

Introduction

The Arab Republic of Egypt is situated in northeastern Africa along the Nile River. Egypt's population of almost 60 million people live in a small portion of Egypt's 386,660 square miles, and this population is expected to rise to over 73 million by the end of this century. Furthermore, much of the land is not arable. The nation experiences a 2.65% annual growth rate in population, but not in food production using current technology. However, through innovative applications of traditional and newer technologies, the growth curves in population growth and food production growth might be made to converge in the early part of the 21st century. In order to achieve this convergence, 150,000 feddans (155,700 acres) must be reclaimed annually.

One technology for improved crop productivity is the use of plastic tunnels as growing environments for young crops. For example, in the nursery, cucumber seeds are started in these plastic tunnels to provide a controlled, protected environment for the development of the seedlings, which are later transplanted to other environments. The tunnel area is 540 square meters. The plastic tunnels afford a measure of protection from draught, insects, wind, and other forces detrimental to the growth of the young, tender plants. The

The projects described in this chapter were funded by the Egyptian Ministry of Agriculture and Land Reclamation (MOALR) and the United Nations Development Programme (UNDP) and were executed by the Food and Agriculture Organization (FAO).

agriculturalist can use these greenhouse-like chambers to eliminate the effects of erosion and to control the amount and quality of water available for the seedlings. The tunnel environment includes these components: plastic tunnels, media, water, seeds, plants, and climate.

These components represent the primary "object classes" in the plastic tunnel network of objects. For each of these components, there are various actions that are performed by the farmers. Operations on the tunnels include tray disinfection and tunnel disinfection. Operations on the media include media preparation, fertilization, salinity reduction, pH reduction, disinfection, and irrigation rescheduling. Operations on the water include salinity reduction and adding pesticides for treatment or protection. Seeds are soaked, plants are sprayed, and the temperature and humidity of the climate are controlled. Figure 7.1 gives an illustration of the plastic tunnel environment.

Expert Systems have the potential to provide intelligent decision support to crop managers, planners, and others where needed to improve the effectiveness and efficiency of crop management decisions. For this particular project, the purposes were to (1) build a local expert system development group capable of planning, development and fielding of expert systems and (2) to develop and package two specific expert systems to assist extension workers and farmers in their goal of maximizing crop output. The first of these systems is designed to provide intelligent assistance for the production of cucumbers raised in protected plastic tunnel environments. This system, termed ESCPMPT/cucumbers (for Expert System for Crop Production Management

FIGURE 7.1 Plastic Tunnel

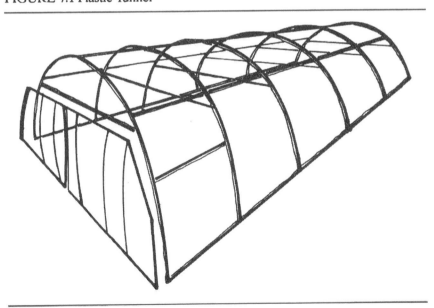

in Plastic Tunnels), is designed to be used by extension agents who provide direct decision support and analysis to farmers at the point of decision making in the fields. (The other expert system is not detailed in this report.)

Knowledge Engineering

Initial Prototyping

Because the first purpose of the project was to build a local expert system development group capable of planning, development and fielding of expert systems, a strong foundation was laid for the development of expert systems. Initially, a technical analysis study to evaluate the operational requirements for this project was conducted. The resulting document (Arab Republic of Egypt 1990b) indicated that the systems should address two main categories of agricultural problems with raising cucumbers and citrus crops. It indicated that they should provide consultation concerning agriculture practices in seedling cultivation, and solution of problems encountered by growers. Such problems are heuristic, non-numeric, and non-algorithmic.

Before beginning any design activities, the project leaders followed a structured knowledge engineering methodology (McGraw, and Harbison-Briggs 1989; Prerau 1985) to select the appropriate agricultural problems, the knowledge engineering team, and other startup activities. Following these activities, an iterative "waterfall" approach for system development was utilized. The prototyping methodology allowed the system development team to test early knowledge acquisition results by building a preliminary system and performing system validation. Further system refinements were based on additional knowledge and feedback from early tests. The resulting system has benefitted from the frequent evaluation and contribution of the domain experts, the project leaders, the knowledge engineers, and users.

Problem Selection and the Knowledge Acquisition Process

This project's initial activity was to determine the specific agricultural problems to be addressed by expert systems. Many agricultural problems could have been targeted. The Ministry officials and project advisors used a structured methodology to select the appropriate decision domain(s) for the project. The project leaders thought that it would be most useful to experiment with the development of a limited expert system which could be applied at the Ministry level. The results of this experiment would be a complete methodology for the production of the operational expert system. Following this experience, an expert system design would provide the basis for further actual computer coding of the expert system. The experiment was designed as follows:

1. Domain experts were recruited from different agriculture sub-disciplines, namely production, nutrition, protection, soils, irrigation, and agricultural economics.

2. Unstructured interviews were held once or twice per week to discuss the project and potential target decisions, and to continue to refine the group's focus.

3. Through these sessions, a limited problem was defined which is cucumber production management at the nursery stage. (Arab Republic of Egypt 1990a)

4. Domain experts were required to write reports explaining their recommendations to growers and what are the problems they frequently encountered when they give advice.

5. These reports were analyzed by the knowledge engineering group and a set of forms were developed to organize conceptually the elicited knowledge.

6. Structured interviews were used to fill out the forms. The initial versions were issued and reviewed. The evaluation and modification resulted in a second version of the forms, which can be seen in Figure 7.2 (in English) and Figure 7.3 (in Arabic).

7. The captured knowledge was intermediately represented using decision trees. A research prototype is scoped as shown in Figure 7.4.

This preliminary design (demonstration prototype) served several purposes. First, it was used as a tool for refining the acquired knowledge. Second, it gave the team leaders an insight into the problem. Third, the experience gained from this prototype was used to develop the requirements specifications and to make a preliminary design for the laboratory prototype (Arab Republic of Egypt 1990d and 1990e).

Availability of expertise, usefulness and potential return on investment of the proposed system, decision complexity, and other factors were considered in the selection of the targets for the knowledge engineering project. The consequence of these initial knowledge acquisition sessions and analyses was to construct an expert system prototype for support in managing cucumber crop production management in plastic tunnels (ESCPMPT).

FIGURE 7.2 Knowledge Acquisition Form

Form-Type-A	Growth Factors at the Nursery Stage of Cucumber	Factor No.
1. Name:		
2. Classification (Environmental/Physiological/Economical):		
3. Determination Method:		
4. Determination Procedure:		
5. Optimal Value:		
6. Recommendation for Reaching the Optimal Value:		
7. Reasons for Deviation from Optimal Value:		
8. Effect of Deviation from Optimal Value:		
9. Correlation with Other Factors:		

Name	Degree	Type
a. _____		

10. Degree of Tolerability: _____

11. Deviation of Handling Procedure:

12. Effect of Deviation Handling on Other Factors:

FIGURE 7.3 Arabic Version of Form

نموذج (أ)	العامل المؤثر في مرحلة المشتل	رقم العامل
١ : اسم العامل		
٢ : تصنيف العامل (بيئي / فسيولوجي / اقتصادي) :		
٣ : كيفية القياس :		
٤ : اجراءات القياس :		
٥ : القيمة المثلى :		
٦ : التوصيات للوصول الى القيمة المثلى :		
٧ : اسباب الانحراف من القيمة المثلى :		
٨ : تأثير الانحراف من القيمة المثلى :		
٩ : ارتباط هذا العامل مع العوامل الاخرى :		

اسم العامل	درجة الارتباط	نوع الارتباط
١ . _____		

١٠ : درجة التحمل :

١١ : اجراءات معالجة الانحراف من القيمة المثلى :

١٢ : اثر اجراءات العلاج من اى من العوامل الاخرى :

FIGURE 7.4. Decision Tree Prototype. A Decision Tree Representing the Knowledge Concerning Disorder Diagnosis for Cucumber Seedlings.

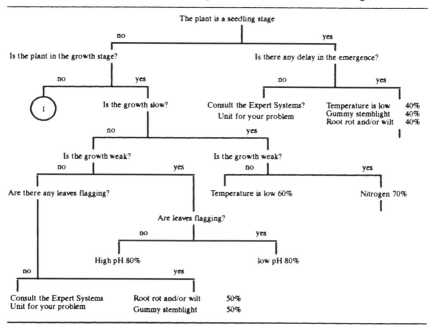

Note: ① represents the start of another whole decision tree.

Feasibility Study

Before proceeding, the team sought to confirm that the planned expert system would be a feasible system. They looked at technical and organizational feasibility, and at various risk and economic issues (Arab Republic of Egypt 1990d and 1990c).

Suitability of the Problem

To evaluate the suitability of the target agricultural subdomain (advise for cucumber production at the nursery stage under plastic tunnels) for expert systems development efforts, the team assessed a number of questions.

Is there an existing algorithmic solution? The preliminary technical report indicated that the system should satisfy two main goals: consultation concerning different issues in cultivations, and consultation concerning solving a problem that a grower may encounter. The nature of such a system deals extensively with non-numeric data and heuristics.

For example, advice for a user who wanted a consultation concerning the preparation of cultivation media would depend on factors such as material available. If all the materials are available, then it might depend on the user's available equipment and tools. Such factors are non-numeric, and lend themselves well to an expert system approach. The rules for such a consultation are depicted in Figure 7.5.

Another example of the heuristic nature of the problem is the certainty factors associated with a conditional clause such as "If plant exhibits yellow leaves." Numerous reasons, such as nitrogen deficiency or low temperatures, could cause this disease symptom. Conventional software is incapable of working effectively with ambiguous data and inexact ("fuzzy") relationships. Therefore, for problems of this nature, developers often choose to apply the technology of knowledge-based expert systems. Expert systems provide acceptable results, advice, or diagnosis even when the available data is ambiguous or incomplete. Expert systems can apply heuristics and inference control strategies to find answers to problems in less time than conventional software, and often find answers when conventional systems can not.

Is the problem of manageable size? As a result of the knowledge acquired and the implementation of the research prototype, it was found that the seedling production management problem is of manageable size.

Does any expertise exist? For cucumber cultivation at the nursery stage under plastic tunnels, there is an expertise that has been accumulated over the last several years. Several experts have been identified in this domain. They

FIGURE 7.5. Decision Rule Examples

RULE NUMBER: 16
If:
    GROWTH MEDIA PREPARATION = 100/100
and Consultation is about: = suitable growth media
and Can you get vermiculite? no
and Containers used are: = trays or paper utensils or plastic bags
and Coarse sand is available
THEN
    the suitable growth media is peatmoss and coarse sand in the ratio 2:1 respectively in volume –

RULE NUMBER: 17
IF:
    GROWTH MEDIA PREPARATION = 100/100
and Consultation is about: = suitable growth media
and Can you get vermiculite? no
and Containers used are: = trays or paper utensils or plastic bags
and Coarse sand is not available
THEN
    the suitable growth media is peatmoss, sand and clay in the ratio 1:1:1 in volume –

RULE NUMBER: 18
IF:
    GROWTH MEDIA PREPARATION = 100/100
and Consultation is about: = fertilizers required
THEN
    FERTILIZERS REQUIRED (for 1 bag peatmoss)
and Magnesium Sulphate (16 gm.)
and Potassium Sulphate (100 gm.)
and Liquid Fertilizer (50 cubic cm.)

---

can articulate their methods for advising growers and assisting them in solving problems.

Does the absence of expertise materially affect the production? Production is most certainly decreased when the identified expertise is not available. During the short cultivation period, if the appropriate decision is not taken, the production will be seriously affected. In some cases, a total crop failure is the result.

Can the knowledge be represented with current technology? In effect, examining the knowledge captured from agricultural experts shows that the nature of the expertise is of the class of knowledge which can be represented, unlike some knowledge classes such as procedural knowledge, physical skills, and highly abstract knowledge. The experts were able to articulate their knowledge in traditional rules and object-oriented constructs.

Characteristics of the Knowledge

Size of the knowledge base. The total overall size of the application is estimated to require over 200 rules. This estimate is based on the feasibility prototype which contained 71 rules and represented about one-third of the overall application. The team concluded that the application could fit the constraints of delivery, especially the hardware constraints. While the Ministry works with more powerful hardware platforms, such as DEC-microvaxes, the goal of this project was to provide deliverable systems which could be literally carried into the field. Implementation on industry-standard laptop computers was a critical factor for project success.

With regard to size, many decision processes (and the related knowledge bases) can be effectively divided into smaller units. These agricultural applications also can be separated into two main modules: one which is concerned with preparation (of materials, media, seeds, etc.) and operations, and the other which is concerned with diagnosis and treatment of problems encountered.

Complexity of the knowledge. Examining the application revealed that it can be divided into small, more manageable modules. For example, knowledge about sterilization of soil can be separated. The same is true for other operations. Diseases and other disorders also can be fit into several general diagnosis problems for which previous work provides solutions.

Time needed to obtain the knowledge. The time needed to develop this application was estimated to be one year. This included all phases of development. Over this period, the knowledge is expected to be fairly stable.

Stability of the knowledge. Most knowledge in this domain is sufficiently stable so that it can be effectively used. In the target application, there is permanent knowledge such as "humidity increases disease probabilities." However, other knowledge, such as the economic variables, is not stable. Relative prices of materials are used to select the more economical materials when there is technical equivalence. This problem can be solved by separating the changeable parameters into a database accessible by the reasoning mechanism of the application.

Form of knowledge. The initial captured knowledge indicated that the knowledge base concerning advice and the cultivation process could be represented in the form of frames. The knowledge concerning disorder analysis could be represented as pattern knowledge that can be implemented using rules.

Characteristics of the Experts

The problem addressed by the project is the development of crop management expert systems to be used by the Egyptian Ministry of Agriculture Extension Offices. Crop management includes different disciplines within agriculture: soil science, water and irrigation, plant nutrition, plant protection, agricultural economics, and production. There are numerous experts in cultivation under plastic tunnels within the Ministry because of the great national interest in protected agriculture. Furthermore, the programme is supported by the UNDP and executed by FAO.

Availability of experts. The identified experts expressed a desire to work on this new project, so their availability was assured. Their interest level remained high throughout the project. Worldwide interest was furthered when the subject was addressed at a workshop held in cooperation with the Siliwood Institute for Pest Management supported by the British Council.

Expert's articulateness. The first interviews with the domain experts revealed that they could articulate effectively their problem-solving methods. They explained their decision rationales and their reasoning processes. This was demonstrated during the development of the feasibility prototype.

Motivation of experts. Those experts who were scientists were motivated by the belief that this new application would help them in their research and teaching activities. Other experts, those working in crop management, learned that this application would help them in their work and also would provide advice for growers. Furthermore, they were financially compensated for their time.

Ministry commitment. Evidence of the Ministry of Agriculture and Land Reclamation's commitment to the project can be seen in their appointment of the First Undersecretary of the Ministry to serve as the National Project Coordinator, who solved problems related to expert recruitment and other management problems. The Ministry also provided space and office facilities.

Expertise dispersal. The knowledge needed for this application needs multiple experts. Overlapping expertise is involved in the application, which increases the complexity of development. However, early experience showed that the experts were able to reach mutual conclusions.

Interfaces

The application requires an interface to provide access to a database which contains materials and material prices. The only other interface that the system will probably use is a graphics presentation package. The system's end users will be precluded from accessing the domain knowledge bases for security reasons.

Prototype Design and Development

Following the initial activities, the team employed an iterative prototyping approach, with early and continual involvement by all team members in system specification, design, and testing. The initial demonstration prototypes were tested and further refined into "laboratory prototypes." These are currently being further validated and refined to become "field prototypes."

The project leaders then developed a plan for the development of these two target expert systems. The initial activities in this plan consisted of:

- knowledge engineering team selection and training
- developing/selecting a knowledge engineering methodology
- hardware and software selection / equipment procurement

To select the primary development tool and development methodolgy for these expert systems, MOALR conducted a study to evaluate alternatives (Arab Republic of Egypt 1990a). The methodology is continuously revised and version 1 of this methodology has been issued (Rafea et. al. 1991b). As a result of the analysis of available tools and their appropriateness for the task, the team decided to use the EXSYS Professional shell, which provides both rule-based knowledge representation and a frame-based representation with inheritance. It also provides blackboarding, agenda management, and a procedural command language deemed to be particularly suited to the team of knowledge engineers. Furthermore, EXSYS Professional provided the team with enhanced graphic interfaces and advanced mathematical capabilities (Rafea et. al. 1991).

Knowledge Representation and Specific Design and Coding

Frames can be defined simply as data structures which describes a concept, an object, or situation. The slots can be defined as the attributes describing the concept, the object, or the situation. A slot can have facets which describe or put constraints on the value to fill the slot.

The method of knowledge representation was based on the objects depicted in Figure 7.6. Each of these primary components in the plastic tunnel environment is related to the other objects in the decision space. Figure 7.7 shows some of the relationships for the media object, along with the slot values for the frames media, raw-material, additions, vermiculite, and peat moss. This object-oriented foundation served as the basis for developing the specific rules used in the ESCPMPT/cucumbers system.

Verification and Validation

The developed system was verified by generating cases from the acquired knowledge. Each case includes a set of descriptors and the decision that should be taken in a situation described by this set of descriptors. A knowledge engineer—not the one who developed the system—takes these cases and runs the system. Discrepancies are marked and necessary actions were taken.

Validation at the laboratory level is done by constructing comprehensive cases which are given to the domain experts. These cases include descriptions of specific situations but the decision is provided by the domain expert. A similar set of these cases is run on the system and the decision is generated by the system. The results are compared and knowledge acquisition sessions are held to discuss these results. The documentation of the verification and validation procedure is being written. The system is now available to be field tested. In September 1991, the expert system will be validated on a daily basis with a real cultivation of cucumbers in a nursery at the Ministry.

Implementation Issues

This section highlights some issues discovered when using EXSYS Professional (El-Dessouki et al. 1991). The frame representation was found to be very weak in EXSYS. Therefore, the design was implemented using the rule representation scheme only. This led to define a procedure to convert frames into rules. During the verification phase, the team noted that there is some redundancy in the questions if the user selects more than one function to perform in one session. To eliminate this redundancy the team decided to use the blackboarding technique supported by EXSYS Professional as a way to pass parameters between linked components of the system.

FIGURE 7.6 Objects in Decision Domain. Overall Domain Structure.

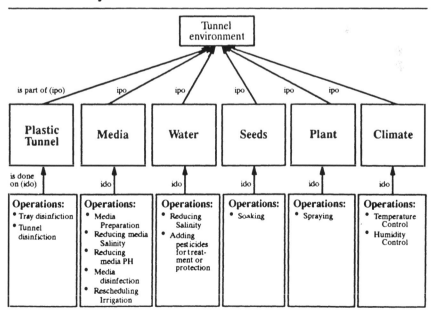

For uncertainty handling the team selected the independent formula supported by EXSYS Professional for several reasons. First, the way in which the knowledge was elicited essentially required experts to respond in terms of IF:THEN structures. Second, EXSYS Professional is reasonably successful at employing certainty factors, an important aspect of the project description.

FIGURE 7.7 The Media Object/Frame (and Slot Values)

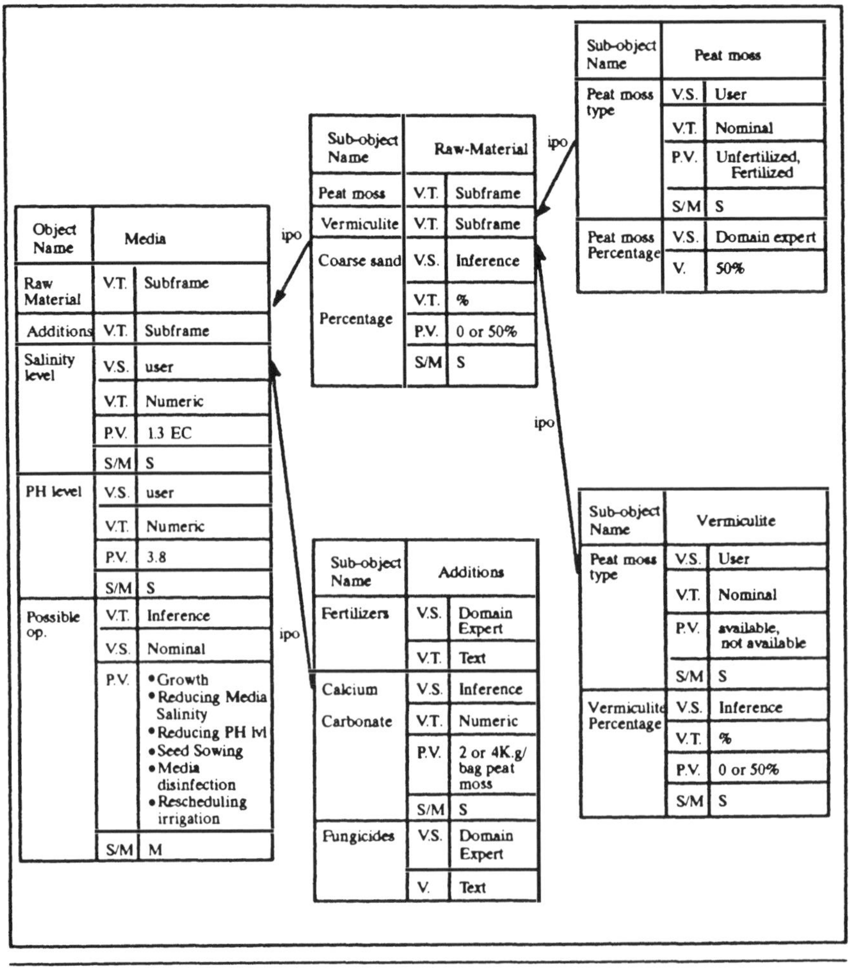

Note: Each slot has a name and facets. The slot itself does not appear in the figure. What appears is the slot name and different facets defined as follows:

V.T.	= Value Type	V.S.	= Value Source
P.V.	= Possible Values	S/M	= Single or Multi Value
S.	= Single	E.C.	= Electric Conductivity

Conclusions and Future Research Implications

Expert systems technology and particularly shell environments have the potential to facilitate more efficient and rapid deployment of major technical innovations, especially where rapid diffusion of innovation is a major goal of national policy, as it is Egypt's agriculture. The current case of cucumber farming in tunnel environments offers an example of the opportunities and the problems in implementing expert systems as a vehicle for technology transfer. On the positive side, the government of Egypt wisely chose to use carefully elaborated techniques of focused domain selection, knowledge engineering options, and selection and evaluation of domain experts and hardware platforms. In addition, the project's leaders went to considerable pains to select the appropriate shell environment and to modify it for their needs. Also noteworthy is the care in planning for validation and verification since many of the reported papers in agricultural expert systems do not effectively address this crucial issue.

On the other hand, it is unclear how successful the project will be in terms of diffusing the technology to field sites. This problem can be addressed through careful beta testing at selected sites. Further, it will be crucial to establish a careful plan for updating the system since all the key parameters of the knowledge involved are relatively volatile from season to season.

Perhaps the most important future implications of this project center around its applicability to other agricultural technology transfer in Africa and Asia. The current system could easily become a test for methodology in selection of domain, knowledge acquisition methodology and verification and validation techniques of relevance to nearly one hundred nations where conventional technology transfer in agriculture is too expensive, too cumbersome, and too delayed to achieving desired effects. It is in the context of this need for a methodology that is solid, replicable and convincingly exportable to developing nations that the current research may make its greatest contribution.

References

Arab Republic of Egypt, Ministry of Agriculture and Land Reclamation. January, 1990a. *Survey of available methodologies of engineering expert systems.* Technical Report TR-88-024-01. Expert Systems for Improved Crop Management.

_____. February, 1990b. *Problem definition of an expert system for cucumber production management at the nursery stage.* Technical Report TR-88-024-02. Expert Systems for Improved Crop Management.

_____. January, 1990c. *Feasibility study for the development of an expert system for cucumber production management at the nursery stage.* Technical Report TR-88-024-03. Expert Systems for Improved Crop Management.

_____. January, 1990d. *Requirement specifications for the prototype of the expert system for cucumber production management at the nursery stage.* Technical Report TR-88-024-05. Expert Systems for Improved Crop Management.

_____. January, 1990e. *Design of the prototype expert system for cucumber production management at the nursery stage-Version 1.0.* Technical Report TR-88-024-06. Expert Systems for Improved Crop Management.

_____. January, 1990f. *Expert Systems for Improved Crop Management,* Technical Report TR-88-024-015.

Buchanan, G.G. 1986. Expert systems: Working systems and the research literature. *Expert Systems.* (3): 32–51.

Caristi, J., A.L. Scharen, E.L. Sharp, and D.C. Sands. December, 1987. Development and preliminary testing of EPINFORM, an expert system for predicting wheat disease epidemics. *Plant Disease.* 1147–1150.

Cooley, D.R. April, 1988. Expert Systems: A new tool for plant pathologists. *Plant Disease.* 279.

El-Dessouki, A., A. Rafea, M. Kamal, H. Safwat. 1991. An expert system for seedling production management, First National Expert Systems and Development Workshop. MOALR, Egypt.

Gevarter, W. May 1987. The nature and evaluation of expert system building tools. *Computer.* 20(5): 24–41.

Halterman, S.T., J.R. Barrett, and M.L. Swearingin. January-February, 1988. Double cropping expert system. *Trans. of the ASAE.* 234–239.

Jones, P., and J. Haldeman. January-February, 1986. Management of a crop research facility with a microcomputer-based expert system. *Trans. of the ASAE.* 235–242.

McGraw, K.L., and K. Harbison-Briggs. 1989. *Knowledge acquisition principles and guidelines.* Englewood Cliffs, NJ: Prentice Hall.

Michalski, R.S., J.H. Davis, V.S. Bisht, and J.B. Sinclair. 1983. A computer-based advisory system for diagnosing soybean diseases in Illinois. *Plant Disease.* (67): 459–463.

Palmer, R.G. September/October, 1986. How expert systems can improve crop production. *Agricultural Engineering.* 28–29.

Peart, R.M., F.S. Zazueta, P. Jones, J.W. Jones, and J.W. Mishoe. May/June, 1986. Expert systems take on three tough agricultural tasks. *Agricultural Engineering.* 8–10.

Prerau, D. Summer, 1985. Selection of the appropriate domain for an expert system. *AI Magazine.* (7)2: 26–31.

Rafea, A., A. El-Dessouki, S. Mohamed, and H. Hassan. 1991a. Knowledge acquisition for crop management expert systems. *Proceedings of the First National Expert Systems and Development Workshop. (ESADW-91)* MOALR, Egypt.

____, A. El-Dessouki, M. Kamal, and H. Safwat. 1991b. An Overview of Four Expert Systems Development Tools. *Proceedings of the First National Expert Systems and Development Workshop. (ESADW-91)* MOALR, Egypt.

Rauscher, H.M. and R. Hacker. Fall, 1989. Overview of artificial intelligence applications in natural resource management. *The Journal of Knowledge Engineering.* 2(3): 30–42.

Ruth, S.R. 1991. Harnessing shell-based knowledge engineering environments: an applications-based perspective. Chapter 2 of this book.

Sinclair, J.B. 1983. A computer-based advisory system for diagnosing soybean diseases in Illinois. *Plant Disease.* (67): 459–463.

Warkentin, M.E., P.K.R. Nair, S.R. Ruth, and K. Sprague. 1990. A knowledge-based expert system for planning and design of agroforestry systems. Chapter 8 of this book.

Waterman, D.A. 1986. *A Guide to Expert Systems.* Reading, MA: Addison-Wesley Publishing Co.

8

A Knowledge-Based Expert System for Planning and Design of Agroforestry Systems

Merrill E. Warkentin, P.K.R. Nair,
Stephen R. Ruth, and Kristopher Sprague

Introduction

This paper describes the preliminary results of the efforts to develop a Knowledge-Based Expert System (KBES) for agroforestry. Known as the United Nations University Agroforestry Expert system (UNU-AES), its goal is to assist individuals in applying the agroforestry approaches to land management for sustainable production of food and fuelwood supplies by farmers in developing countries.

While this book describes numerous applications of expert systems to agriculture, we are not aware of any attempts to apply this approach to agroforestry. We believe that it could represent a significant step in the planning and design of agroforestry systems.

Agroforestry and Alley Cropping

Agroforestry is a collective name for land use systems in which woody perennials are deliberately grown on the same piece of land as agricultural

The project was financed by the United Nations University, Tokyo. Mr. Michael Bannister of the Department of Forestry, University of Florida, and Dr. Dennis V. Johnson of the USDA Forestry Support program, Washington, D.C., worked with the domain expert (Nair) and provided substantial inputs to develop the system. The knowledge engineers (Warkentin, Ruth, and Sprague) were assisted by several of their colleagues at George Mason University, notably Dr. Ken De Jong and Mr. Michael Wollowski of the Computer Science Department.

A longer version of this chapter was published originally in *Agroforestry Systems*, Vol. 11, 1990, pp. 71–83. Copyright 1990, Kluwer Academic Publishers. Reprinted by permission of Kluwer Academic Publishers.

crops and/or animals either in some form of spatial arrangements or in sequence. In agroforestry systems, the woody component interacts ecologically and economically with the crop and/or animal components. Such interactions take many different forms, both positive and negative, and they need not remain stable over time. The aim and rationale of most agroforestry systems are to optimize the positive interactions in order to obtain a higher total, a more diversified and/or a more sustainable production from the available resources than is possible with other forms of land use under prevailing ecological, technological, and socio-economic conditions (ICRAF 1982). There are several examples of indigenous agroforestry systems in the developing countries of the tropics and subtropics (Nair 1989).

Alley cropping is a type of agroforestry system in which leguminous trees are planted in rows with food crops cultivated between them. The trees are pruned on a regular basis to minimize shade to associated crops, provide nitrogen-rich foliage, recycle nutrients, and, in some cases, to provide additional outputs such as fodder and firewood. As the tree foliage may serve both as mulch and fodder, the system also provides a means of integrating crops and animal production. Economic analyses have indicated alley cropping may be a profitable enterprise for farmers in certain environmental conditions. Thus, alley cropping is of considerable interest to researchers, policy makers and farmers in developing countries. A substantial body of research information on various aspects of alley cropping from different parts of the tropics has become available during the past five years, (Attah-Krah and Francis 1987; Budelman 1988a; 1988b; Duguma, Kang, and Okali 1988; Gill and Patil 1985; Huxley 1986; Jama and Nair 1989; Jama et al. 1986; Kang, Wilson, and Lawson 1984a; 1984b; Kang, Grimme, and Lawson 1985; Kang and Duguma 1985; Kang and Wilson 1987; Nair 1984; Nair 1987; Sumberg 1986; Wilson, Kang, and Mulongoy 1986; Yamoah, Agboola, and Wilson, 1986a; Young 1987).

Development of the Agroforestry Expert System

The process of developing knowledge-based expert systems, known as knowledge engineering, includes several distinct steps. In developing the UNU–AES, the methodologies proposed by Harmon and King (1985), Waterman (1986), Weitzel and Kerschberg (1989), and others were followed. Typical phases of expert system development are:

1. Selection of an Appropriate Problem
2. Development of a Prototype System
3. Development of a Complete System
4. Evaluation of System
5. Integration of a System
6. Maintenance

Alley cropping was selected as the focus of this first effort in the application of expert system technique to agroforestry because of the availability of knowledge in forms appropriate to expert system construction, and because of its potential applicability to the practice of agroforestry practice in developing countries.

Waterman (1986) suggests that within the development of a prototype system, there are actually three prototype systems developed. The first is the demonstration prototype, the purpose of which is to determine if the approach ("proof of concept") is viable and system development is achievable. The second prototype, known as the research prototype, displays credible performance on the entire problem. In most cases, the system is "fragile" in several areas due to incomplete testing and revision. The third is the field prototype, the purpose of which is to display good performance with adequate reliability of the overall system. This objective is achieved by extensive testing in the user environment (Waterman 1986). This stage of development precedes the development of a complete system and is the current status of UNU-AES.

Specifying Performance Criteria

The performance criterion selected for this project was to have UNU-AES achieve results that were consistent with those generated by the primary domain expert. In other words, each time UNU-AES would be consulted, it was expected to provide the same advice as a recognized expert on alley cropping. Although this is quite optimistic, it established a benchmark against which to test UNU-AES.

Selecting the Development Environment and Tools

EXSYS Professional, by EXSYS, Inc., was selected as the expert system building tool to develop UNU-AES. Several criteria guided the selection of this software package. Because of the expected demonstration and use of the expert system in developing countries, the first requirement was that the software run under the MS-DOS operating system and that it operate on a portable computer. Secondly, it was important that the program be simple enough so that an end user could learn to run the program and perform sensitivity analyses with minimum amount of training time. The third consideration was the wealth of powerful features of the software tool. Chief among these features were its screen definition language (to customize the way questions are asked) and its ability to allow interface calls to external programs. EXSYS Professional also provides an explanation facility to assist the system user.

Turbo Pascal, by Borland, Inc., was selected as the software package to create the screen graphics. This software package was chosen because it fulfilled the operating system and target computer requirements, it could be accessed from EXSYS via external program calls, and it offered the developers a set of tools to develop the color screens.

Developing a Lab Prototype

In the development of the initial version of UNU-AES, several key decisions regarding the components which make up an EXSYS knowledge base were addressed, including the selection of a probability system. This system provides the range of probability values which can be assigned to each rule in the knowledge base. To make results easy to interpret, the knowledge engineers selected a probability system which provided four groupings between 0 and + 100. The probability values are based on four classes indicating varying degrees of success that could be expected when the recommended technology is implemented. The four probability classes are:

poor – 20/100 (range of 0–39)
fair – 50/100 (range of 40–59)
good – 75/100 (range of 60–85)
excellent – 90/100 (range of 86–100)

Other terms used in connection with the system are defined below: (Sprague and Ruth 1988)

Qualifier — Contains a key determinant and appears in the IF statement in the form of a question. Most qualifiers usually end with a verb.

Qualifier Value — The possible completion of the IF statement. A qualifier value may be a number, a word, or a phrase.

Choice — Appears in the THEN statement and is the possible solution to the decision rule. A choice has a probability value associated with it.

Probability Value — Is a numerical value assigned to the choice and indicates the likelihood of successfully implementing the recommendation.

Decision Rule — Is the conditional statement which consists of facts and/ or heuristics about the specified area.

Note — Information which tells the user who, what, where, when, why, and how something occurs.

System Description and Operation

UNU-AES will be available on a microcomputer floppy disk. When the system is run on an MS–DOS microcomputer, the first image that appears on the monitor is the subject screen, which contains information on the expert system and the individuals who developed it. After viewing this screen, the user may press any key to see the next screen that will provide an overview of the system and a brief explanation on how to run the UNU-AES. The next screens ask the user questions (qualifiers) regarding rainfall, altitude, and soil of the subject location, and so forth. For practical reasons, the number of qualifiers, as well as the number of options under each qualifier, is limited in this first version of UNU-AES, as shown in Figure 8.1. For each qualifier, the user may select the class that describes the situation under consideration as closely and appropriately as possible, and give the answer by pressing the appropriate number key. When the user is finished answering the last text screen, any key may be pressed to see the results on the screen. Based on the information that has been provided, UNU-AES will give a set of outputs that includes one or more of the five tree (hedgerow) species considered in this program, hedgerow spacing for each, and an associated probability (confidence) value. As discussed above, the probability values are based on four classes indicating varying degrees of success that could be expected when recommended technology is implemented. The user has the option to have all of the choices contained in the results displayed, or only the choices which exceed some predefined threshold value. The results also can be printed out. The choices for hedgerow species and row spacings are given in Figure 8.2. An illustrative decision rule is given in Figure 8.3. An example of sample outputs as they appear on the computer screen is given in Figure 8.4.

FIGURE 8.1 UNU-AES System Qualifiers (Basis for Questions Asked of User)

No.	Qualifiers	a	b	c	d
1.	Annual rainfall (mm)	700–1099	1100–1600	> 1600	
2.	Number of rainy seasons/yr	1 (unimodal)	2 (bimodal)		
3.	Total rainy months/yr	3–5	5–7	7–9	> 9
4.	Elevation (m)	< 500	500–1000	1000+	
5.	Slope (%)	< 15	15–29	30–39	> 40
6.	Soil texture	Clay	Loam	Sandy	
7.	Soil fertility	Poor	Medium		
8.	Soil reaction (pH)	< 5.5	5.5–7.5	> 7.5	

FIGURE 8.2 UNU-AES System Choices (Resulting System Advice Options)

Plant hedgerows of:

Leucaena leucocephala (Lele) spaced
    < 4 m , 4 to 6 m apart, or > 6 m or greater apart.
Leucaena diversifolia (Ledi) spaced
    < 4 m , 4 to 6 m apart, or > 6 m or greater apart.
Gliricidia sepium (Glse) spaced
    < 4 m , 4 to 6 m apart, or > 6 m or greater apart.
Calliandra calothyrsus (Caca) spaced
    < 4 m , 4 to 6 m apart, or > 6 m or greater apart.
Cassia siamea (Casi) spaced
    < 4 m , 4 to 6 m apart, or > 6 m or greater apart.

FIGURE 8.3 A UNU-AES Decision Rule

IF:	The approximate annual rainfall per year is low (700 to 1100 m.)
and	The number of rainy seasons in a year is one (unimodal)
and	The average number of rainy months in a year is three to five
and	The elevation/altitude of the field is (low) less than 500 m.
and	The predominant soil reaction is acidic (less than 5.5 pH)
THEN:	
	Plant Leucaena leucocephala (Lele) spaced less than 4 m. apart or spaced 4 to 6 m. apart or spaced 6m. or greater apart.– Confidence = 20/100
and	Plant Leucaena diversifolia (Ledi) spaced less than 4 m. apart or spaced 4 to 6 m. apart or spaced 6m. or greater apart.– Confidence = 20/100
and	Plant Cassia siamea (Casi) spaced less than 4 m. apart. Confidence = 20/100
and	Plant Cassia siamea (Casi) spaced 4 to 6 m. apart or spaced 6m. or greater apart.– Confidence = 50/100

FIGURE 8.4 UNU-AES System Output (Sample Advice Given by AES)

	Values based on 0 to +100 system	*VALUE*
1	Plant Leucaena leucocephala (Lele) spaced less than 4 m apart or spaced 4 to 6 m. apart.	90
2	Plant Leucaena diversifolia (Ledi) spaced less than 4 m. apart or spaced 4 to 6 m. apart.	90
3	Plant Gliricidia sepium (Glse) spaced less than 4 m. apart or spaced 4 to 6 m. apart.	90
4	Plant Calliandria callothyrsus (Caca) spaced less than 4 m. apart or spaced 4 to 6 m. apart.	90
5	Plant Leucaena leucocephala (Lele) spaced 6 m. or greater apart.	75
6	Plant Leucaena diversifolia (Ledi) spaced 6 m. or greater apart.	75
7	Plant Cassia siamea (Casi) spaced less than 4 m. apart.	75
8	Plant Gliricidia sepium (Glse) spaced 6 m. or greater apart.	75
9	Plant Calliandria callothryrsus (Caca) spaced 6 m. or greater apart.	75
10	Plant Cassia siamea (Casi) spaced 4 to 6 m. apart or spaced 6 m. or greater apart.	50

Conclusion

The system in its present form has several limitations, but there are immense possibilities for improvement by adding more qualifiers and by broadening the choices under each qualifier. Data and knowledge pertaining to socio-economic characteristics also should and can be built into the system to make it more useful. The system could also be extended to cover other agroforestry technologies such as improved fallows, plantation crop combinations, homegardens, soil conservation hedges, and so forth. The effort also serves to bring to the attention of agroforestry research scientists the type of field data that will need to be generated in order to use such data with decision-making tools such as knowledge-based expert systems.

References

Atta-Krah, A.N., and P.A. Francis. 1987. The role of on-farm trials in the evaluation of composite technologies: the case of alley farming in southern Nigeria. *Agricultural Systems*. 23:133–152.

_____ and G.D. Kolawole. 1987. Establishment and growth of *Leucaena* and *gliricidia* alley cropped with pepper and sorghum. *Leucaena Research Reports*. 8:46–47.

Budelman, A. 1988a. The decomposition of the leaf mulches of *Leucaena leucocephala*, *Gliricidia sepium*, and *Flemingia macrophylla* under humid tropical conditions. *Agroforestry Systems*. 7:33–45.

_____. 1986b. Leaf dry matter productivity of three selected perennial leguminous species in humid tropical Ivory Coast. *Agroforestry Systems*. 7:47–62.

Carrol, J.M., and J. McKendree. 1987. Interface design issues for advice-giving expert systems. *Communications of the ACM*. 30:14–31.

Caristi, J., et al. 1987. Development and preliminary testing of EPINFORM, and expert system for predicting wheat disease epidemics. *Plant Disease*. 71: 1147–1150.

Colfer, C.J.P., R. Yost, F. Agus, and S. Evensen. Expert systems: A possible link from field work to policy in FSR. Photocopy.

Duguma, B., B.T. Kang, and D.U.U. Okali. 1988. Effect of pruning intensities of three woody leguminous species grown in alley cropping with maize and cowpea. *Agroforestry Systems*. 6: 19–35.

Gill, A.S., and B.D. Patil. 1985. Intercropping studies with *Leucaena* under an intensive fodder production programme. *Leucaena Research Reports*. 5:36–37.

Halterman, S.T., et al. 1988. Double Cropping expert system. *Trans. of the ASAE*. 31:234–239.

Harbans, L., R.M. Peart, and J.W. Jones. 1988. FARMSYS – A knowledge-based (decision) support system for multi-crop farming systems.

Presented at the 8th Annual Farming Systems Research / Extension Symposium. October 9–12, 1988.

Harmon, P., and D. King. 1985. *Expert Systems: Artificial Intelligence in Business.* New York: John Wiley & Sons, Inc.

_____, R. Maus, and W. Morrissey. 1988. *Expert Systems: Tools and Applications.* New York: John Wiley & Sons, Inc.

Huxley, P.A. 1986. Rationalizing research on hedgerow intercropping—an overview. Nairobi: ICRAF Working Paper 40.

ICRAF (International Council for Research in Agroforestry). 1982. What is agroforestry? *Agroforestry Systems.* 1: 7–12.

Jama, B. A., Getuhan, D. Ngugi, and B. Macklin. 1986. *Leucaena* Alley cropping for the Kenya Coast. In *Amelioration of Soils by Trees.* ed. R.T. Prinsely and M.J. Swift. London: Commonwealth Secretariat.

Jama, B. and P.K.R. Nair. 1989. Effect of cutting heights of *L. leucocephala* hedges on production of seeds and green leaf manure at Machakos, Kenya. *Agroforestry Systems.*

Kang, B.T., and B. Duguma. 1985. Nitrogen management in alley cropping systems. In *Nitrogen in Farming Systems in the Humid and Sub-humid Tropics.* ed., Kang and Van der Heide. Haren, Netherlands: Institute of Soil Fertility.

Kang, B.T., H. Grimme, and T.L. Lawson. 1985. Alley cropping sequentially cropped maize and cowpea with *Leucaena* on sandy soil in southern Nigeria. *Plant and Soil.* 85: 267–277.

Kang, B.T. and G.F. Wilson. 1987. The development of alley cropping as a promising agroforestry technology. In *Agroforestry: A Decade of Development.* ed. H.A. Steppler and P.K.R. Nair. Nairobi: ICRAF.

Kang, B.T., G.F. Wilson, and T.L. Lawson. 1984a. Alley cropping maize (*Zea mays* L.) and *Leucaena* (*leucaena, Leucocephala Lam*) in Southern Nigeria. *Plant and Soil.* 63: 165–179.

_____, 1984b. *Alley cropping: A stable alternative to shifting cultivation.* Ibadan: IITA.

Nair, P.K.R. 1984. *Soil productivity aspects of agroforestry.* Nairobi: ICRAF.

Nair, P.K.R. 1987. The ICRAF field station, Machakos: A demonstration and training site for agroforestry technologies. *Agroforestry Systems.* 5: 383–394.

Nair, P.K.R., ed. 1989. *Agroforestry Systems in the Tropics.* Dordrecht, The Netherlands: Kluwer Acad. Publ.

Ruth, S.R. and C.K. Carlson. 1988. Shell-based expert systems in business: A return on investment perspective. *Proceedings of the Ninth International Conference on Information Systems.*

Sprague, K.G. and S.R. Ruth. 1988. *Developing expert systems using EXSYS.* Santa Cruz, CA: Mitchell Publishing Co.

Sumberg, J.E. 1986. Alley cropping with *Gliricidia sepium:* Germplasm evaluation and planting density trial. *Tropical Agriculture.* 63: 170–172.

Warkentin, M.E. 1988. Expert system shells. *Decision Line.* 19: 9–11.

Waterman, D.A. 1986. *A Guide to Expert Systems.* Reading, MA: Addison-Wesley Publishing Co.

Weitzel, J.R. and L. Kerschberg. 1989. Developing knowledge-based systems: Reorganizing the system development life cycle. *Communications of the ACM.* 32: 482–488.

Wilson, G.F., B.T. Kang, and K. Mulongoy. 1986. Alley cropping: Trees as a source of green manure and mulch in the tropics. *Biol. Agriculture and Horticulture.* 3: 251–267.

Yamoah, C.F., A.A. Agboola, and K. Mulongoy. 1986. Decomposition, nitrogen release and weed control by prunings of selected alley cropping shrubs. *Agroforestry Systems.* 4: 239–246.

Yamoah, C.F., A.A. Agboola, and G.F. Wilson. 1986. Nutrient contribution and maize performance in alley cropping systems. *Agroforestry Systems.* 4: 247–254.

Young, A. 1987. *The potential of agroforestry for soil conservation; Part II: Maintenance of fertility.* ICRAF Working Paper 43. Nairobi: ICRAF.

9

Expert Systems for Information Transfer About Soil and Crop Management in Developing Countries

*Russell Yost, Stephen Itoga, Zhi-Cheng Li,
Carol Colfer, Le Istiqlal Amien,
Phoebe Kilham, and James Hanson*

Introduction

The goal of the research in Indonesia is to develop soil management principles and practices appropriate for the large number of "transmigrants," persons encouraged to migrate from highly-populated Java to the less-populated outer islands. Because conditions are much different on these islands, many of the farmers from Java do not have experience in farming the acid uplands so prevalent there. For this reason, our research has been oriented toward helping the transmigrants manage highly acid soils.

An important part of our research has been the use of expert systems as a medium to apply information transferred from analogous areas of the tropics,

The expert system development effort described in this chapter is part of the TropSoils research program at the University of Hawaii. Tropsoils is a project funded by the Agency for International Development in which four universities are developing soil management technology to permit more efficient use of soil and crop resources in the humid tropics (Peru, North Carolina State University; Indonesia, University of Hawaii); acid savannas (Brazil, Cornell University), and the semi-arid tropics (Niger, Texas A&M University). The TropSoils research described here was conducted by the University of Hawaii; the Centre for Soils Research, Bogor, Indonesia; and, for a period of time, North Carolina State University. An interdisciplinary Tropsoils staff has been doing intensive fieldwork in rural Sumatra. At the time much of the data described in this paper was collected, the staff consisted of soil scientists, agronomists, an anthropologist, a nutritionist, and a graduate student in economics.

to apply results of project research within the country to benefit directly the people there, and to apply our research results in ways to benefit other areas. Our expert system work began nearly five years ago with a prototype for determining lime requirements for managing soil acidity. This led to the development of the three expert systems we describe in this paper: FARMSYS, ADSS (Acidity Decision Support System), and LIMEAID.

Combining Local Information with International Expertise

Preliminary soil surveys of the research site in the Sitiung transmigration area of West Sumatra indicated that the soils were similar in taxonomy to acid soils from other countries of the tropics (Table 9.1). Many of the problems associated with farming these soils already had been faced and management practices had been developed to deal with the acidity. The procedure used by the soil acidity expert system for calculating lime requirements was first developed in these regions.

The project staff oriented their research toward solving transmigrant farmers' problems rather than purely academic problems. Principles and procedures resulting from the research might then be applicable to other areas of the humid tropics. The staff used research and development strategies described by Shaner et al. (1982) to ensure that the soil and crop management practices were relevant to the transmigrant farmers. This strategy included collecting survey information concerning the farm family and using the survey results in planning project research. Transmigrant farm and farm family characteristics were also important considerations when applying soil management experience from other countries. For example, the high soil acidity led agronomists to search for aluminum tolerant crops. They suggested growing cowpea rather than mungbean. However, mungbean is a preferred

TABLE 9.1 Countries with Soils Analogous to Those Identified at the Tropsoils Project Site in Sitiung, W. Sumatra

Soil Great Group	Countries Where This Great Group is Found
Humitropepts	Kenya, Ecuador, Guatemala, Costa Rica, Philippines
Dystropepts	Kenya
Tropudults	Philippines, Brunei
Palehumults	Kenya, Thailand, Chile, Papua New Guinea, Philippines
Paleudults	Rwanda, Peru
Haplohumox	Lesotho, New Zealand, Brunei
Haplorthox	Thailand, Philippines, Papua New Guinea, Costa Rica, Rwanda, Brazil

Source: Soil Conservation Service and Soil Management Support Services.

food among the transmigrants—even if it means extra lime or animal manure applications, they would rather grow mungbean. The agronomically optimum crop was not optimum when local preferences were considered.

In an earlier paper (Colfer et al. 1989), we described the experience of trying to use an expert system framework to capture socio-cultural research findings from the project in Sumatra. Local farmer information is essential for understanding local conditions and for designing appropriate technology and problem-solving strategies. The expert system FARMSYS is an exploratory attempt to make more useful some of the information on indigenous and transmigrant farmers. The goal of the FARMSYS system is to integrate information about the farmers and the farm family into a system that provides recommendations about agricultural research from a farm household's perspective. This system is intended to help agricultural researchers examine potential implications of new research objectives. The anthropologist on the project had an important role in helping plan field research, including collaborative farmer research.

In the course of the fieldwork, it became evident that that were a diversity of cultural views about soil. For example, the indigenous people, mostly Minangkabau, prefer low-labor agriculture while the Javanese and Sudanese transmigrants take pride in tilling and caring for the soil and the crops they grow on it. The Javanese and Sudanese are not averse to being advised to hoe lime into their land. The hoe is part of their culture. For the Minangkabau, however, the opposite is true—they much prefer to grow rubber, which grows well on acid soils, because so little work is required to maintain the crop. Even in their language the Minangkabau equate acid soil with good soil while the Javanese and Sudanese consider acid soil poor soil.

The knowledge we try to capture in FARMSYS includes information that is difficult to translate into the relatively rigid goal-oriented structure of the rule-based expert system. We are not sure that the knowledge representation method (rules and text) we are using is appropriate for the knowledge we are trying to capture. Despite these difficulties, however, we believe there are several advantages to the expert system approach: (1) The system can handle large quantities of qualitative and quantitative data and incorporate this information into reasoning links if such reasoning exists. (2) The system provides a mechanism whereby someone from another discipline can get a near "expert opinion" on a question without having to learn the methods of that discipline. (3) This type of system can provide an avenue for making village-based information available to planners and decision-makers.

There are at least two approaches to dealing with a diversity of sociocultural information. One is to identify principles in the information so that likely behavior can be inferred given selected characteristics. Another method is to attempt to develop a knowledge base that can be partitioned into modules and loaded into a decision-support system. Until principles can be extracted from experience, considerable detail must be stored in the knowledge base and new situations must be dealt with by analogy or by similarity to existing information.

Interdisciplinary knowledge of the local situation is required for developing strategies for dealing with soil acidity. Socioeconomic factors should be considered along with crop and soil factors. FARMSYS represents an attempt to make more explicit this socioeconomic dimension. Crop and soil management factors are dealt with in a separate expert system, ADSS. However, before describing that system in detail, we first note some of the benefits we have observed in the course of working with this expert system approach to formulating soil and crop management recommendations.

Benefits of Expert Systems for Soil Acidity Management

One of the greatest benefits of the expert system approach is that it permits knowledge and expertise from several disciplines to be used in evaluating a problem and in devising a solution. The system does not ask difficult questions unless the user is provided with alternative ways of answering. Much more work needs to be done on how to gather and represent information from various disciplines so that the information can interact in a harmonious manner. An intriguing thought is that interdisciplinary interaction may be easier to develop in expert systems than in teams of experts. Could this be a way to realize some of the outstanding advantages of interdisciplinary problem-solving?

Our soil acidity management systems are still being developed but have already demonstrated a potential for organizing information, both mathematical and experiential, into a decision-making framework and generating sound recommendations. The systems are easy to use and can assist the user in learning elements of the decision-making process. The process of developing the expert system has challenged us to learn, to organize information and to deal with uncertainty. As a result of this process, direct information retrieval has been gradually shifting from microprocessor-based technology.

One of the advantages of the expert system is its ability to record and put into use more than one type of knowledge. Both heuristic knowledge acquired through experience and algorithmic knowledge gained through traditional research efforts can be represented in the knowledge base. Experiential and nonquantitative information have often been considered of secondary quality. However, in trying to capture an expert's thought and rationale for a particular recommendation, both types of information are needed.

In agriculture, we often underestimate the importance of blending heuristic and algorithmic information. Heuristic information can augment rigorous information generated as a result of well-designed and well-conducted experiments. Experiential information can provide an alternative method of solving of a problem when a quantitative solution is not known or is not available. All agricultural information should be tested in realistic situations. For example, the introduction of highly intensive agriculture requiring careful hoeing and tilling of the soil is likely to fail if the farmers are Minangkabau while chances of success are great if the farmers are Javanese or Sudanese transmigrants.

Early Development

Our efforts began with the expectation that expert systems would enable us to record a procedure for determining lime requirements that had been tried and found useful in areas analogous to the study site in West Sumatra. The system was expected to serve as a transfer mechanism to and from environmentally similar areas of the tropics. We hoped that the system would contribute to scientific effort by clearly and objectly laying out the procedure for determining lime requirements for crops growing on acid soils of the humid tropics. This open description of the procedure, sometimes called 'transparency', was also thought to be an excellent way to teach the lime recommendation methodology to our research staff. In addition, we had ambitious plans for using expert systems to capture and preserve soil management and farm family information for later use.

With these aspirations, we began to work on an expert system for managing acid upland soils of the humid tropics. The soil acidity prototype did not actually leave the development computer, fortunately. The first system to show promise was ACID2—a system that did minimal lime requirement computations and only made recommendations of lime required in intervals, e.g. 1 to 1.5 tons/hectare. This preliminary system first asked for the soil classification and the type of crop. With these two pieces of information, the system then asked for soil chemical information (extractable aluminum, ECEC, and soil pH), checked these values for consistency, and then calculated the lime requirement using the equation proposed by Cochrane et al. (1980), which takes crop tolerance to aluminum saturation into consideration. ACID2 explained its logic by displaying the rules (sequences of IF ... THEN ... clauses) used in the reasoning. The subsequent expert system, ACID3B, was the first to employ external graphics routines, to do computations using external procedures written in PASCAL, and to give internal estimates of consistency for most variables entered. ACID3B was distributed to over 100 scientists who had cooperated in the initial effort and who had expressed interest in testing the system for us.

The need to retrieve data from an external soil database and to include an economic analysis motivated us to develop the ACID4 expert system, subsequently updated and renamed ADSS (Acidity Decision Support System). ADSS is designed for a wide audience. As only general information is asked for, the user need not be a soil scientist. Specific information is contained within the expert system itself. For example, the user is asked only to select a region and a soil profile typical of that region and ADSS retrieves from the soil database technical information on the soil profile including extractable aluminum, effective cation exchange capacity, and soil pH. ADSS is described in more detail in the following section.

The ADSS Expert System

ADSS is based upon the four main modules of ACID4: (1) A soil database containing soil pH, effective cation exchange capacity (ECEC), extractable aluminum, and bulk density at three depths for about 250 representative soils; (2) A list of crops and their tolerance to soil acidity (critical aluminum saturation); (3) A soil management module that determines factors such as depth of lime incorporation, amount of organic material added, quality of the lime, and levels of potassium and phosphorus; and (4) An economics module that performs a partial-budget analysis of the economic impact of liming on crop profitability using such factors as lime cost, transport cost, application cost, duration of liming effect, and opportunity cost of funds used for purchasing lime.

The soil database contains data required for estimating the amount of lime necessary to reduce soil acidity by neutralizing soil aluminum. The database contains about 200 soil pedons from Sitiung, other areas of Indonesia, Thailand, Malaysia, the Philippines, and several islands in the Pacific Basin. Users can determine lime requirements, management recommendations, and profit based on soils representative of large regions.

The soil database module allows the user to select the country and soil series whose classification and soil chemical data will be used in the recommendation. Effective cation exchange capacity, extractable aluminum, soil pH, and bulk density values will be passed to the expert system for use in developing lime recommendations. These values will be considered default values to be used unless the user wishes to enter other data instead. After exiting the database, some or all of the data entered can be changed and the system rerun. This permits the user to see how sensitive the lime recommendation is to individual components of the data. From this, the user can determine which kinds of data are the most critical to the recommendation.

After the program exits the soil database, the user must choose the type of crop for which the recommendation is to be generated. Information on the relative tolerance of the various crops to soil acidity is critical for effective crop management on acid soils. This information was determined from published reports of acidity tolerance of crops and from research conducted by project staff with local cultivars of soybean, rice, peanut, and cowpea (Wade and Al-Jabri 1984). The aluminum tolerance of the crop is inferred from the crop selection. Aluminum tolerance is the level of aluminum saturation the soil will be adjusted to by liming, if liming is necessary. A typical cropping season includes two crops during the rainy season. Effects of liming on both crops and associated economic consequences are calculated. Crop management information also includes recommendations for crop sequences representative of local practices. If the computer has a graphics display, a graph of the estimated relative yield will be shown.

In the soil management module, various factors that can modify the lime requirement are determined. Soil management information can be obtained either from typical soils of the region or from specific analysis of a single soil sample. The amount of lime needed to reduce soil aluminum saturation to a level that will not reduce crop yield is estimated using an equation developed by Cochrane et al. (1980). These include consideration of the depth of liming with cautions generated if the intended depth of incorporation is either too shallow or too deep. If the crop is aluminum sensitive, the aluminum saturation of the soil is high, and the incorporation depth is too shallow (less than 15 cm), the system warns of possible drought due to shallow rooting and consequently reduced amounts of available water (Gonzalez et al. 1979). The system will identify situations in which these conditions are so severe that crop failure is likely. Crop failure may be caused by retardation of root growth and function by aluminum. If root growth is limited to the surface 15 cm of soil it is unlikely that the crop will survive more than a few days without rain. The expert system can easily flag these situations and warn the user of the risk. This module is currently being expanded to describe more adequately the severity of drought resulting from shallow rooting.

Addition of organic material is another factor that may lead to modification of the lime requirement. Organic material additions seem to enhance crop yields during drought years (TropSoils/Indonesia 1987). Although yield increases due to organic material are difficult to quantify, it has been shown that organic acids modify the toxic effect of aluminum (Hue, Craddock, and Adams 1986). However, yield increases may also be due to additional nutrients from organic material brought in from areas external to the fields. Studies are needed to determine the nature of this benefit so that it can be managed more effectively. The increase in yield due to organic material additions is an example of information not sufficiently understood to be described by a mathematical equation, yet is information that should be considered in making lime recommendations as it could lead to a reduction in the amount of lime required.

When lime is recommended, questions are asked about both the chemical and physical qualities of the lime. The amount of lime recommended can be adjusted according to the neutralization of acid relative to calcium carbonate and its expected reactivity as influenced by fineness.

Several checks are made to see if data presented to the system are consistent with expectations. If soil pH is less than 4.5 and there is virtually no extractable aluminum, the system will alert the user that this is unusual and the data should be verified. The system will point out that soils of kaolinitic mineralogy usually would have considerable aluminum at pH 4.5. Similarly, if there is aluminum at pH's greater than 6.0, the system will warn the user that this is unusual and that the data from which the system was developed did not have this characteristic. The aluminum-based recommendation methodology used by ADSS does not diagnose soil acidity problems for all mineralogical groups. For example, in volcanic ash soils aluminum toxicity is usually not the most important yield-limiting aspect of soil acidity. Yields may be limited by calcium deficiency, phosphorus deficiency, or manganese toxicity (Fox et al. 1986).

If the Sitiung region is selected, additional options are available including a relay cropping pattern option. When two crops are grown in sequence, both crops need to be considered in making the lime recommendation. If the crops have widely differing tolerances to aluminum, lime will be either wasted on one crop or applied in insufficient quantities to the other crop. The selection of crops for relay cropping patterns is monitored by ADSS and a caution is given if a large amount of lime is required by one crop but not by the other. At the end of the liming assessment portion of the questioning, additional questions are asked relating to soil test levels of phosphorus and potassium. If data are available, recommendations will be provided that are based on information from the project's field studies in Sitiung.

Economic analysis is a relatively recent module. Estimated profit or loss from liming is an important consideration in deciding whether to lime. We presently use a partial-budget analysis that includes the cost of lime, transportation to the farmer's home and to his field, and the costs of spreading the lime. If farmers routinely hoe their fields, the cost of lime incorporation is not included. The value of the potential increase in crop yield due to the addition of lime is estimated from farmer or extension agent estimates of high and low yields. The various costs are then subtracted from the expected increased value of the crop due to the liming. We also report profit or loss as a percentage of investment in lime. Parts of the economic analysis module are tailored to subsistence farming conditions where funds for investment are scarce and ideal recommendations cannot be carried out (Kahn 1987). We have found that considering the residual value of the lime can reduce costs by as much as one third depending on the expected duration of the residual effect. These results suggest that more research to evaluate the residual effects of lime is justified.

Analysis of weather data has not yet been incorporated into the expert system. In soils of the Oxisol and Ultisol orders, it is important to consider weather data when evaluating the effect of lime in ameliorating the root-restricting effects of toxic aluminum. Even though applying lime to highly acid soils can enhance the tolerance of crops to drought, there has been little study of the importance of this effect relative to the direct effects of lime on aluminum toxicity, calcium availability, or pH. As suggested earlier, shallow incorporation of lime can restrict the supply of water available to the crop. The experience of organizing the knowledge and using the knowledge base in the expert system has emphasized the need to estimate more accurately the effects of crop, soil, and lime application on crop water relations. This experience has renewed our interest in including a water balance model in the system to estimate more accurately water stress effects. The process of organizing liming information into a single diagnostic system also brought out other effects on the soil-crop-climate-farmer system that need to be considered.

Computer Implementation of Knowledge Base

Having assembled information from many sources, we were concerned about how to represent accurately the information in an expert system. We wanted to represent the information in a manner that would facilitate decision making and would place minimal restrictions on the hardware required to run the system. We initially designed the system for users who had specific soil data and wanted to know how to interpret their data for a lime recommendation. Soon, however, it became apparent that many liming policy decisions were made by agricultural planners who did not have access to our knowledge base nor to the specific data needed to run the expert system. Our latest system, ADSS, is designed to make information from our project and an international database available in a user-friendly microcomputer software program.

There may be users who are interested in trying the expert system but are not aware of some of its limitations. When a requested task is outside its expertise, the system specifies that the knowledge base only applies to certain crops and soils. If the user chooses a soil group for which the use of this methodology is not recommended (e.g. Andept, a suborder of soils including volcanic ash soils), the system terminates early and provides the user with a literature reference for further information. This sort of checking is done both on soil data transferred to the system from the soil database and on soil data entered by the user.

The Soil Conservation Service and the Soil Management Support Service have provided us with a soil database comprised of several hundred pedons. From this main database we have extracted a minimum dataset of soil information, "filtering" it to remove inconsistent data. For completeness, there are soils in the dataset for which our expert system does not currently contain the correct lime requirement methodology. This dataset together with soil survey information from Sitiung makes up the soil database. Included in this database are data for the major Southeast Asian countries, the Pacific Basin, and several countries in Africa. If another country or region is selected, the data format is sufficiently simple that the soil data can be quickly processed and substituted for data currently in the system.

We began implementing the soil acidity knowledge base by organizing knowledge traditionally used to determine lime recommendations. Information was sorted from general (soil data, type of crop, tolerance of the crop to soil acidity, etc.) to more specific (calculation of amount of lime needed to reduce soil acidity to levels tolerated by the crop). This information was then recorded using rules within an expert system shell (EXSYS 1990). An expert system shell is a preprogrammed structure that facilitates entering information into an expert system. A shell provides a mechanism for recording information but contains no information itself. It usually consists of editors and tools to build "rules". Rules are statements made of IF ... THEN ... sequences that simulate rational thought and deduction. The following is an example of a rule: IF a farmer lives in village X THEN it is likely that the farmer is ethnic group Y. Various levels of uncertainty can also be attached to the rule to reflect the certainty of the statement. Arithmetic computations can also be made in rules. In this way rules can suggest specific conclusions as well as calculate specific quantities (such as amounts of lime or fertilizer for recommendations).

One advantage of using a shell at the beginning to implement a knowledge base is that a prototype expert system can be developed very quickly. In our case we were able to develop a prototype system within three to four weeks. This included the basic structure of two of the modules (crop requirements and soil management). After developing the prototype we distributed it to three to four TropSoils colleagues with whom we discussed our first impressions of the expert system. These discussions gave rise to many suggestions and led to modifications that were implemented within a few days. This pattern of testing, discussing, and modifying the system was continually repeated over the next two years. With time, the circle of testers became larger and the discussions and modifications more detailed. The TropSoils field staff was included in all discussions. As a result of this testing process, the need for an international soil database and for an economics module became apparent.

An expert system shell is designed for the early stages of building expert systems because it makes it easy to write rules, change them around, and add, modify, or delete as necessary. These kinds of changes are far less easy in a programming language. However, the expert system shell can eventually get in the way of the expert system development as more detailed refinements are made. As the program becomes more complex, a programming language can provide more flexibility in allowing the program to perform as the knowledge requires. Indeed, one of the tenets of software development is that the programming language (or programming environment) should not impede the job to be done.

Although the ADSS expert system is written using a shell as the basic framework, to increase flexibility, we also began developing another expert system, LIMEAID, using the declarative language Prolog to implement the knowledge base. We found that Prolog is more appropriate for writing the advanced form of the knowledge base (Arity 1988). Prolog is often used in logic programming and in expert system development. It is considered a logic programming language because it uses logic to make inferences from data. Programming with Prolog is almost like stating what should be done rather than having to instruct the computer exactly how to do it. Most of the ADSS knowledge base was also implemented in LIMEAID.

Representing the knowledge base in a different software environment gave us a new perspective on how best to record the knowledge and to interact with the user. The system of menus and prompts in the programming language environment of LIMEAID is more user friendly—the user need only choose from menus or enter data if the default values are not desired. In addition, the declarative style led us to consider ways we could use the knowledge base to answer new types of questions. For example, can we use the knowledge base to determine which soils should be used given constraining crop options, economic factors, and soil management? Similarly can we determine the best crops given soil, economic, and soil management constraints? Implementing the knowledge base in a different expert system format has also helped us distinguish between the knowledge and the implementation structure. Developing the LIMEAID system, however, came at the cost of requiring a trained Prolog programmer to write the program and to maintain it.

Conclusion

In this chapter we have described an attempt to use expert systems methodology to organize knowledge about managing acid soils and to transfer that knowledge to those making agricultural planning decisions. The problem-solving knowledge is highly interdisciplinary and is composed of both internationally available information as well as local farmer and site-specific information. Our expert system ADSS can be used for interpreting site-specific data for local landuse planning and for accessing an international soil database for large area landuse planning.

Recently educators have been stressing the value of thinking about how we think (metacognition). Our work in expert system compelled us to think about how we make recommendations and how we develop and manage new research information. Therefore not only has the ES itself proved useful in landuse planning, but the process of developing it has led to a better knowledge base.

The information gathering and organizing techniques we have described should be applicable to many different types of development projects. Ensuring local adaptability—with on-site surveys of farmer conditions and on-farm research—was essential for full and appropriate use of international information. The resulting appropriate practice could, in turn, both be applied locally and extrapolated to analogous regions of the tropics. Although as yet inadequate to represent all the information generated by interdisciplinary research projects, these methods provide exciting new ways to capture and transfer problem-solving knowledge.

References

Arity Corporation. 1988. *Arity/Prolog Programming Language*. Concord, MA.

Cochrane, T.T., J.G. Salinas, and P. Sanchez. 1980. An equation for liming acid mineral soils to compensate crop aluminum tolerance. *Trop. Agric.* (Trinidad). 57:133–140.

Colfer, C.J.P., R. Yost, F. Agus, and S. Evensen. 1989. Expert systems: a possible link from fieldwork to policy. In *Farming Systems*. AI Applications in Natural Resource Management. 3:31–40.

EXSYS, Inc. 1990. *EXSYS: Expert System Development Package*. Albuquerque, NM.

Fox, R.L., R.S. Yost, N.A. Saidy, and B.T. Kang. 1986. Nutritional complexities associated with pH variables in humid tropical soils. *Soil Sci. Soc. Amer. J.* 49:1475–1480.

Gonzalez-Erico, E., E.J. Kamprath, G.C. Naderman and W.V. Soares. 1979. Effect of depth of lime incorporation on the growth of corn on an Oxisol of central Brasil. *Soil Sci. Soc. Am. J.* 43:1155–1158.

Hue, N.V., G.R. Craddock, and F. Adams. 1986. Effect of organic acids on aluminum toxicity in subsoils. *Soil Sci. Soc. Am. J.* 50:28–34.

Kahn, S. 1987. Observations on the limited resource transmigrant farmers of Sitiung 5C, West Sumatra. Preliminary Report. Gainesville, FL: Department of Food and Resource Economics, University of Florida.

Shaner, W.W., P.F. Philipp, and W.R. Schmehl. 1982. *Farming Systems Research and Development: Guidelines for Developing Countries*. Boulder, CO: Westview Press.

TropSoils/Indonesia. 1987. *TropSoils Technical Report* 1985–1986. Raleigh, NC: TropSoils Management Entity, North Carolina State University.

Wade, M. and Al-Jabri. 1984. Lime efficiency and prediction for soybean on acid upland soils of Indonesia. TropSoils/Indonesia. Unpublished Manuscript.

10

DREAGIS: A Knowledge-Based Agricultural Geographic Information System for the Dominican Republic

*David Mendez Emilien
and Severin V. Grabski*

Introduction

The economy of the Dominican Republic is heavily dependent on agricultural production, therefore good information for agricultural planning purposes is essential. This chapter describes a prototype knowledge based agricultural geographic information system for use in the Dominican Republic (Dominican Republic Expert Agricultural Geographic Information System DREAGIS). The system is designed to aid agricultural planning by: identifying crops that are suitable for a given set of soil and climatic conditions; displaying, on maps developed through a geographic information system, where those crops can be grown without regard to current land use (i.e., what can grow in this area regardless of the current plantings); again displaying, on maps, where those crops can be grown after taking into consideration current land use (i.e., where can this crop be grown without disrupting current plantings).

The chapter is organized as follows. We first describe the characteristics of a geographic information system, and differentiate it from mapping and database management systems. We then identify some tools currently used in agricultural planning, including some expert systems that are under development. Next, we describe the particular geographic information system that we employ, then the current planning environment in the Dominican Republic. We also discuss why a knowledge-based geographic information system was chosen for this environment. We then describe the implementation details and present future work that needs to be accomplished.

Geographic Information Systems

A Geographic Information System (GIS) is a computer program or a set of computer programs designed to store, edit, manipulate and retrieve data that can be geographically referenced (Burrough 1986; Cowen 1988). In other words, a GIS stores maps of a region in computer files for further processing. The objective is that for a given geographical area, several maps can be stored in a GIS attending to different attributes of the area under study. Such attributes, or themes, depend on the type of analysis that is going to be performed. Typical examples of maps used in a land evaluation task include: soil type, climate, rainfall level, contours, current land use, slopes, population density, rivers, and infrastructure. Each map records a single attribute or theme of the region under study.

The maps are introduced into the computer storage facilities in several ways, geocoding, digitizing, scanning, and direct input from video output. In geocoding, a grid is overlaid on each map, dividing it into a certain number of rows and columns. The rectangular area that is formed by the intersection of a row and a column is called a "grid cell," and it is considered to be homogeneous in terms of the attribute being analyzed. Each grid cell on a map is uniquely identified by the corresponding numbers of its defining row and column. The topic or theme represented in the map is then coded according to the number of pre-specified levels. Each grid cell then, contains one and only one code, representing the level of the attribute under consideration. If more than one level of the attribute coexists within the same grid cell, different procedures may be used to select one as the representative of the entire cell; one of the most popular procedures is to select the level with the largest area within the cell. For example, if a cell contains both land and water, the cell would be coded as water (given, say, a code value of 1) if the water area is over 50% of the cell area. If the land area was greater than 50%, then the cell would be coded as land (given, say, a code value of 2).

For each map or attribute, a computer file is created, and each record in the file consists of the row number, column number, and attribute level of a specific cell. This procedure can be repeated for different maps to obtain a set of computer files, each one storing a different attribute of the same geographic region. The grid cells must be consistent from map to map for any comparison among maps or combination of maps to be possible.

A digitizer can be used to input the coordinates and attribute values directly from the map to the computer. Alternatively, the map can be scanned in, or it can be directly input via video (either NTSC or RGB) output into a special "frame-grabber" display board (Miller et al. 1989). Once the maps of the region have been stored in files, the GIS allows the combination of two or more attributes in response to queries. This response can be generated in the form of numerical tables or as a new map. The response can be displayed on a high resolution screen, or it can be printed or plotted. It is this automated creation of a database that helps distinguish a GIS from a database management system or a mapping system. Additionally, the ability of the GIS to synthesize existing levels of geographic data and update a database of spatial entities is a key feature that helps define a GIS (Cowen 1988).

A GIS can be used to develop maps to show the proximity of growing areas to cities and roads for transportation. It can also be used to help identify regions to grow various crops, and to help determine the appropriate government agency or agencies that will be required to organize the work force if the agricultural and political regions are not isomorphic. Typical queries to the GIS for a land evaluation task might be:

- Compute the area of forest coverage in the region;
- Find all the areas with an elevation less than 1000 feet, slope between 0 and 8% and that are located within 5 miles of a major road (conditions that might be suitable for growing a certain crop), and compute the area of the subregion defined by the matching cells and show the subregion in a scaled map;
- Compare the current and potential land use maps. Show in a composed map, the subregions that are being "well" used and those in which the current land use does not match the potential use, and compute both areas.

There are, however, inherent drawbacks in a GIS, mainly because all the data must be discrete and geographically identifiable. Questions such as the kind of agricultural inputs used in a certain operation or the physical conditions necessary to grow a certain crop cannot be answered by the GIS.[1] To overcome these limitations, a standard database system must be used.

The Comprehensive Resource Inventory and Evaluation System Geographic Information System (CRIES-GIS)

Currently, some of the agricultural information for the Dominican Republic is stored in a Geographic Information System that was developed by the Comprehensive Resource Inventory and Evaluation System (CRIES) project at Michigan State University (Schultink et al. 1987).

As a part of the CRIES project, the agricultural regions of the Dominican Republic were classified in Resource Planning Units (RPUs). The concept of an RPU is to have each unit be defined according to identifiable geographic and climatic conditions. For example, an RPU of type 5 could be found in the

---

[1] The GIS provides information about current soil and geophysical conditions. It does not provide information about what can grow under these conditions. These characteristics must be supplied through an external database.

western part of the country and also in the southern part. A crop that is currently growing in the western part would be expected to have similar success in the southern part. The data contained in the CRIES-GIS for the Dominican Republic is based on this concept of RPUs.

The CRIES-GIS is a full function GIS providing such features as statistical analyses, data analyses including erosion modeling and proximity analysis, and terrain analyses including slope characteristics and spatial filtering (Schultink et al. 1987). It is one of the few GISs designed to operate in a microcomputer environment using the MS-DOS operating system (current implementations exist in systems using the Intel 8088, 80286, and 80386 microprocessors). Many other GIS that operate in a microcomputer environment require the addition of expensive boards, and the cost for a complete system can easily exceed $80,000 for the hardware and software. In contrast, the CRIES-GIS functions in a microcomputer environment, does not need any special cards, and the software license and necessary hardware (including digitizer, high resolution video, and ink jet printer) can be obtained for less than $15,000. This low cost and access to "commodity type" computer hardware is important in developing countries where there is not a large budget for capital acquisitions, and general purpose, multiuse equipment is preferred to specialized hardware.

Agricultural Planning Needs in the Dominican Republic

The planning officials of the Agriculture Ministry of the Dominican Republic specified the objectives for the agricultural information system. They want to be able to identify what kind of crops are being grown in any site of the country (total area, total production, and yield). Additionally, they want to be able to identify where a specific crop is being grown and whether or not the sites of production have the physical conditions required for the crop. Ministry employees often receive questions from farmers as to what crops can be grown on their farm (other than the current plantings). They also receive questions from potential investors as to what locations of the country are best suited for the farming of a certain crop for which the price is rising on the international market. Officials believe that they must develop a better way to address these requests. Currently, there is not any established procedure to answer these questions, and the individuals who possess the needed expertise are scattered throughout the Ministry. Ministry officials also want to cross-reference files of areas with files of crops to determine the "best" spatial distribution of crops based on the yield, cost of production, or profit. Finally, they would like to identify the specific training required by the agricultural extension technicians according to the crops produced in different areas.

Answers to some of the Ministry officials' questions can be provided through the CRIES-GIS. However, the problem we face with the CRIES-GIS (and with any GIS) is that it does not contain all the data required by Ministry officials. It cannot be queried, say, to find out the names and training of the agricultural extension technicians that work in a specific area, nor can it be queried to determine what crop will grow in a certain area (this requires expertise in specific plant species). This type of data, however, could be readily available in a database environment.

We considered implementing only a conventional database package, but in response to a query, such a database would not be able to produce a precise geographical definition of an area in form of a map. The ability to provide responses in map form was considered to be a critical factor for the successful use of this system for agricultural planning. Consequently, the solution adopted was to link the CRIES-GIS with a database package, and to employ an expert system to provide expertise in identifying the crops that could be grown in a particular area. An expert system is appropriate for situations where heuristics and symbolic reasoning are employed rather than algorithmic approaches, and for situations that are bounded and are knowledge intensive (Bobrow et al. 1986; Harmon and King 1985; Ruth and Carlson 1988; Waterman 1986). It is also appropriate for situations in which the expertise must be disseminated across a population which has a wide variance in the skills embodied in the expert system.

Use of Geographic Information Systems in Conjunction with Expert Systems

Because it incorporates the use of both an external data base and the use of a GIS this expert system differs from other expert systems that have been developed in the area of agricultural planning, such as PEANUT (developed to aid in irrigation decisions for a particular type of peanut in a specific geographic region) (Ruth and Carlson 1988), COMAX (developed to aid in cotton farming and harvesting decisions) (Lemmon 1986), or AES (developed to support land use involving alley cropping under defined conditions in the tropics and subtropics) (Warkentin et al. Chapter 8) DREAGIS also is more extensive than these planning tools because it is designed for a multi-product environment, rather than focusing on a single crop or type of crop.

Expert systems have been used in conjunction with geographic information systems in a variety of settings. Robinson et al. (1987) classified expert systems for use with GIS into four areas: map design; geographic feature extraction; geographic database systems; and geographic decision support systems. Map design systems were developed to improve map quality for GIS derived maps (MAP-AID (Robinson and Jackson 1985), for placement of feature names on maps (AUTONAP, Freeman and Ahn 1984), to emulate a cartographer in map generalization (MAPEX (Nickerson and Freeman 1986), and for atlas design (CES (Muller et al. 1986). Geographic feature extraction systems such as ACRONYM, (Brooks 1983), FES, (Goldberg et al. 1984), and others utilize the expert system to interpret aerial and satellite images.

Geographic database systems have been developed to search through very large geographic databases to find sets of spatial locations which satisfy a query or to find all spatial objects that exist in a specific area (KBGIS-II, Smith et al. 1987). Geographic decision support systems (GDSS) generally incorporate related non-geographic factors, including economic and/or social criteria. GDSS have been developed for use in aspen management (ASPENEX, Morse 1987), for regional planning in which the requirements of the economic sectors (e.g., agriculture, industry, mining, etc.) and social requirements (e.g., living conditions, employment, education, etc.) are compared (Barath and Futo 1984), and to help planners evaluate location suitability for particular land-use activities (GEODEX, Chandra and Goran 1986).

Our prototype expert GIS can be classified as a geographic database system. The current objective is to aid in the planning of land use based on crop and environmental characteristics. The system indicates whether certain controllable factors of production are possible (e.g., irrigation is a potential factor to control for lack of precipitation, however, the slope and drainage must be within certain bounds) in order to expand the list of potential crops that can be grown in an area. The long-range objective is to expand the system so that it will incorporate the costs of the factors of production and social planning along the lines of Barath and Futo (1984). The system could then be classified as a geographic decision support system.

Explanation of the DREAGIS System

Our explanation of the system will first present how the database portion interacts with the CRIES-GIS. Next we explain how the knowledge-based system is employed. The database portion of the system was constructed using dBase IV, and it consists of two main modules, data storage and editing, and queries. The database query capability, when combined with the GIS, will allow Ministry officials to identify what crops are being grown in any site in the country. The knowledge-based portion of the system was constructed using VP-Expert. This module will allow Ministry officials to identify what crops can be grown in a certain region, the locations best suited for the farming of a certain crop, and whether current sites of production have the agrophysical conditions required for good growth. When combined with the GIS, this module will allow Ministry officials to identify what crops can be grown in a certain region after taking into consideration the current land use.

Database

A menu system (Figure 10.1, Panel A) is used for data storage, editing, and printing reports about the crops (Figure 10.1, Panel B) and regions (Figure 10.1, Panel C). The database can be queried using standard dBase IV query commands. A menu system has not been developed for any database queries due to the varied nature of the queries. It is possible that a menu may be added at a later point if a standard set of queries are identified.

The region table in the database contains data about the regions which are categorized according to RPUs. At this time, our prototype only contains data about the Ocoa Watershed in the Dominican Republic and crops grown in the Ocoa Watershed. It also contains the digitizing code of each region, and this is used to link the regions to the GIS. For each region the physical factors considered to be most relevant for crop development are recorded. These factors are: pH, precipitation, temperature, drainage, soil depth, salinity, and slope (Laureano and Amparo 1985). The crop table in the database contains the same factors, and it also contains the crop classification. Both tables have almost identical structures. For each of the physical characteristics listed, a range is stored in the database for every crop and region. In the crop table this range represents the set of values required for the crop to grow in any given region. In the region table the range represents the span of the distribution of the values that the specific physical characteristic attains in that region. In Figure 10.2 we present an overview of the crop and region tables. In Figure 10.3, Panel A, presents the information stored about a particular crop—onions— and Figure 10.3, Panel B contains the information stored about region 02A (digitizing code 1).

FIGURE 10.1 Database Menu System

```
┌─────────────────────────────────────┐
│ Agroecological Information System    │
│    Database Maintenance Menu         │
├─────────────────────────────────────┤
│          1. Crops Menu               │
│          2. Regions Menu             │
│                                      │
│          0. Quit                     │
└─────────────────────────────────────┘
```

Panel A. Main Database Menu

```
┌──────────────────────────┐   ┌──────────────────────────┐
│    Crop Specification     │   │    Region Specification   │
├──────────────────────────┤   ├──────────────────────────┤
│ 1. Add Information        │   │ 1. Add Information        │
│ 2. Change Information     │   │ 2. Change Information     │
│ 3. Remove Information     │   │ 3. Remove Information     │
│ 4. Review Information     │   │ 4. Review Information     │
│ 5. Print Information      │   │ 5. Print Information      │
│                           │   │                           │
│ 0. Exit                   │   │ 0. Exit                   │
└──────────────────────────┘   └──────────────────────────┘
```

Panel B. Crops Database Menu   Panel C. Regions Database Menu

FIGURE 10.2 Crop and Region Data Types

DREAGIS SYSTEM Crop and Region Database	
CROP TABLE Fields: pH Precipitation Temperature Drainage Soil Depth Salinity Slope	**REGION TABLE** Fields: pH Precipitation Temperature Drainage Soil Depth Salinity Slope

FIGURE 10.3 Crop and Region Data

CROP ID NUMBER	5	NAME	ONIONS
CLASSIFICATION	*HORTICULT*	*MIN*	*MAX*
pH	= = >	4.1	8.3
Precipitation	= = >	500	2500
Temperature	= = >	6.0	30.0
Drainage	= = >	2.0	6.0
Soil Depth	= = >	25.0	1000.0
Salinity	= = >	0.0	2.0
Slope	= = >	0.0	30.0

Panel A. Crop Data

REGION ID NUMBER 1		NAME	02A
		MIN	*MAX*
pH	= = >	5.5	7.0
Precipitation	= = >	1400	2000
Temperature	= = >	25.0	25.0
Drainage	= = >	6.0	7.0
Soil Depth	= = >	25.0	50.0
Salinity	= = >	0.0	2.0
Slope	= = >	30.0	100.0

Panel B. Region Data

Knowledge-Based System

The knowledge-base component of this system was developed using the VP-Expert software. The intent is to provide as much capability as possible while using standard products that have a reasonable cost (many of the works referenced require mini-computers and specialized software). The VP-Expert software also has the capability to interface directly with the dBase IV data files. This was also a critical factor in our decision to use this software. In Figure 10.4 we present an overview of the DREAGIS system components. The expert system shell reads and writes directly to the database files. When using this system, a user will actually be querying the database as to current land usage, and so on. A conversion program, DB2GIS, is then used to generate data in a format compatible with the CRIES-GIS. This data is then read into the GIS and the appropriate maps are generated. The system is bi-directional, meaning that a digitizer can be used with the GIS to determine, say location, and this information can then be converted from GIS format into the database format, and then be used in a consultation with the knowledge-base system (this is the method that will be used when a farmer wants to know what crops a specific site is capable of supporting). The current implementation does not support the execution of the conversion program from within the expert system. However, we plan on adding this feature and hence show a broken line between the conversion program box and the expert system.

FIGURE 10.4. DREAGIS Components

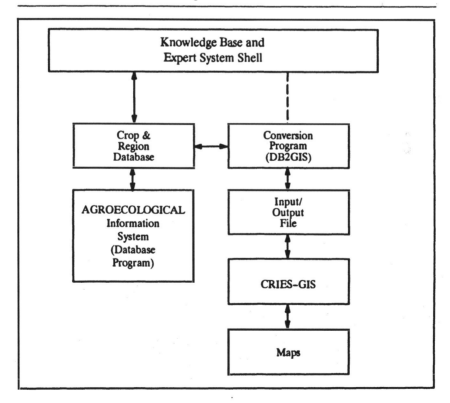

The procedure employed by the knowledge base component involves the following process. For a given crop, each physical factor is compared to the equivalent factor in a potential region. If, for a factor, the range of the region is completely contained within the range of the crop, then the crop can grow in that region based on that factor. Otherwise the region is not suitable for the development of that crop, it is rejected, and the next region is evaluated. If a region satisfies all the factors for a crop, the region is regarded as suitable for the development of that crop. In the future, we plan on relaxing this constraint. So that if, for a factor, the range is not completely contained within the range of the crop, but at least some overlap exists, the system will classify the region as suitable for the development of that crop regarding the specific factors, but with a warning that the adaptability does not exist for the entire region. In order to implement this feature, we will need to take into consideration the constraints that are critical for the growth of a certain crop, (e.g., minimum temperature or maximum salinity), and not allow inclusion of the area if it fails to meet the critical factors. Currently, if there is the possibility of irrigation, the precipitation factor can be relaxed. However, in order to irrigate, the slope of the region cannot exceed 15 percent. The system automatically checks the slope constraint if irrigation is identified as a possibility. Figure 10.5 presents an overview of this process.

Our prototype knowledge base system opens with a menu screen which provides the user with a choice of tasks (Figure 10.6). The following tasks are currently supported:

- Identification of where a specific crop is currently grown;
- Identification of where a crop could be grown based upon that crop's specific agrophysical characteristics;
- Identification of where a crop could be grown based upon its similarity to other crops; and
- Identification of what crops can be grown in a specific farm or RPU.

We also plan to include a fifth menu item, identification of where a crop could be grown with an expected good yield based on the similarity of the "good yield" area to other areas in the country. This has not been implemented in this prototype due to limited data availability. The four listed items were implemented because they provide a mechanism to disseminate the expertise that was deemed most needed at this time, the identification of where crops can be grown, and where crops are currently grown. Due to the lack of established procedures to answer these questions, and since the individuals who possess the needed expertise are scattered throughout the Ministry, it has been difficult for employees to perform this task in a consistent manner. Officials believe that they must be able to perform this task in a credible manner before any other agricultural planning suggestions can be successfully implemented. This system will provide Ministry officials a means of consistent dissemination of this information.

FIGURE 10.5 DREAGIS System Region Selection Criteria

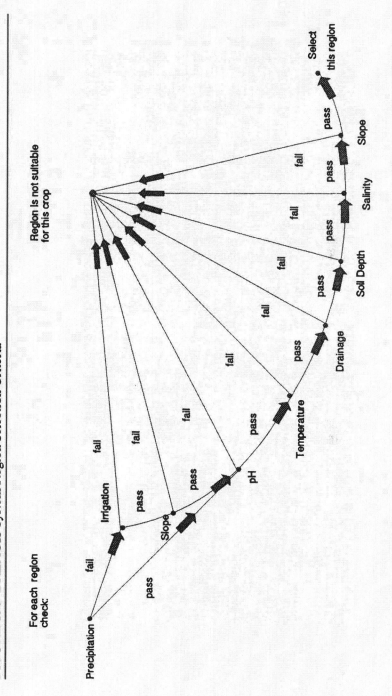

After a user has decided on a task, the knowledge-based system will query the user for any additional information that is needed. For example, if a user wants to identify where a crop could be grown based on similar crop characteristics, the user is first asked to identify a similar crop, and a list of crops is presented (Figure 10.7). In the prototype, we have implemented five basic crop groups: potatoes, coffee, onions, bananas, and beans. The crop is then "typed" or classified by the knowledge-based system, i.e., as a perennial, musoceal, and so on. At this point the system presents a listing of all crops within the selected plant family. From that listing, the user can select the crop that most closely matches the characteristics of the desired crop (Figure 10.8). After this choice is made, a list of crop agrophysical characteristics is displayed (Figure 10.9). The user is then provided the opportunity to modify any of the characteristics, say amount of rain, and also identify whether irrigation will be a possibility (Figure 10.10). The revised list of characteristics is displayed for the user to verify. When the user is satisfied that the list is correct, the knowledge-based system then references the database for all regions that meet the specified characteristics (Figure 10.11).

The output resulting from the consultation session is then linked to the GIS through the conversion program, DB2GIS (which is written in the computer language C). DB2GIS converts the database table that was created as a result of the consultation session to one that cross-references the areas in the table with the areas of maps of areas stored in the GIS. As a result, DB2GIS will produce a file-map for the GIS where all the areas contained in the table will have an attribute area of 1 and all others a value of 2. With this information in the GIS, the user can display or print a scale map of the selected areas and perform other GIS relevant analysis procedures such as proximity to cities, proximity to rivers, slope, or any other desired output. In Figure 10.12 we present the map of the Ocoa Watershed in the Dominican Republic produced by the CRIES-GIS, and in Figure 10.13, we present the areas of the watershed that supports the crop growth, that is, the map of the areas listed in Figure 10.11.

FIGURE 10.6 Knowledge-Base Menu

What type of crop decision do you want to make?
1. Where a crop is currently grown.
2. Where a crop could grow based on crop characteristics.
3. Where a crop could grow based on similarity to other crops.
4. What crops can be grown in a specific RPU.

FIGURE 10.7 Crop Identification

Is the considered crop similar to: ?
    Potato
    Banana
    Beans
    Coffee
    Onions

FIGURE 10.8 Crop Selection

I have ONIONS classified as a
HORTICULT crop. Within the group
HORTICULT, is your crop similar to: ?
    Pepper
    Garlic
    Onions

FIGURE 10.9 Crop Agrophysical Characteristics Screen

The physical characteristics of ONIONS are:

ONIONS	MIN	MAX
pH	4.1	8.3
Precipitation (mm)	500	2500
Temperature (C)	6.0	30.0
D ainage	2.0	6.0
Soil Depth (cm)	25.0	
Salinity	0.0	2.0
Slope (%)	0.0	30.0

Press any key to continue

FIGURE 10.10 Irrigation Divisions

Are you considering i rigation?
    No   Yes

FIGURE 10.11 List of Feasible Areas

The following is a list of the regions where
this crop can be grown:

Region	Digitizing No.
40C	10
40B	0
02C	3
02B	2

FIGURE 10.12 Map of Ocoa Watershed

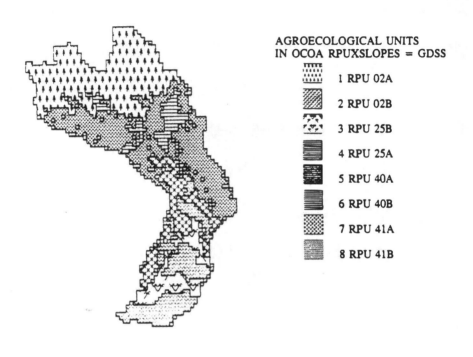

AGROECOLOGICAL UNITS
IN OCOA RPUXSLOPES = GDSS

1 RPU 02A

2 RPU 02B

3 RPU 25B

4 RPU 25A

5 RPU 40A

6 RPU 40B

7 RPU 41A

8 RPU 41B

The Dominican Republic
Agricultural Information System

Current Situation

The Ministry of Agriculture of the Dominican Republic is currently using an Agricultural Information System (AIS) in a GIS environment to store and analyze data relevant to the development and implementation of land use policies at the national level. Even though specific goals have been developed for the use of the system, the scope of its impact in the organization transcends these goals. The AIS has become a framework for the activities of several policy-making units in the Ministry of Agriculture, and they have redefined their goals around the AIS.

Currently, different units of the Ministry are in charge of the design of the data structure, data input, data maintenance, modeling, and information retrieval. The Agricultural Information System has lead to an increased

FIGURE 10.13 Areas of Ocoa Watershed that Support Onions

XOCOAGRP.RAS

RPUs SUITABLE FOR ONIONS
IN THE OCOA WATERSHED

1 AREA SUITABLE FOR
ONIONS

2 AREA NOT SUITABLE
FOR ONIONS

awareness among the planners for a well defined geographic structure that can be represented with minimal cost and effort in the data base. This structure has been designed to match the resolution of the information needed to implement policies at a national level. At this time, the major task consists of the definition of homogeneous areas of the country that can be treated as single entities regarding physical condition and policy implementation. The fact that the political division of the agricultural extension service does not match the physical division of homogeneous agroecological units in the country leads to the requirement to have both reference frameworks in the database in order to perform policy analyses using the physical framework for the analysis and the political framework to implement the policies. Based on this perspective, the system has been a major success. It has shown the disparity between the two structures and as a result, forces have begun to re-shape the political divisions and make them more compatible with the physical areas. At the same time, the information system has provided a bridge between planning and implementation, areas that used to be largely uncoordinated.

For policy evaluation, the agricultural information system is currently used in a "what if" mode. The next step is to link the set of primary data concerning the performance of implemented policies to procedures that will lead to an improvement on the current situation of land use at the national and regional levels. This will be achieved via two different simultaneous approaches, an expert system approach and a procedural approach. While neither approach has been implemented, the plan is to have the expert system accumulate facts about the current land use, policy implementations and the results of such policies. Then, when a problem is posed, the expert system will try to match the conditions to some previously analyzed problem and provide recommendations of what to do (or what not to do) based on similarities with the previous case. In the event that the conditions of the new problem are totally new to the system, a theoretical base will be necessary for the system to combine the facts, and the system will utilize the theory base to derive a set of "close-to-optimal" policies. These will be submitted to the decision makers for evaluation, and once implemented, will be stored as part of the experiences of the system.

Planned System Enhancements

The current implementation only considers agrophysical and geographical characteristics in the determination of whether a crop can grow in a given area. While we allow for the relaxation of the precipitation constraint through irrigation, we do not allow other factors to be relaxed. As discussed previously, this will be changed in the future. The current implementation does not take into consideration any socio-economic factors. Since an objective of this project is to allow Ministry officials the best information for decision making regarding what crop should be grown in a given area, these factors must be included.

Four steps are required for the determination of what crop should be grown in a given area. These include feasibility analysis, cost/benefit analysis, optimization, and sensitivity analysis. At its present state of development, this system only addresses the feasibility analysis aspect of what crop can be grown based on agrophysical and geographic constraints. Consequently, the next step will address the cost/benefit issues. This requires the collection of costs of production and estimates of crop prices. As with any benefit/cost analysis, the objective will be to determine whether it is profitable to grow the crop. The optimization and sensitivity analysis will be the third aspect of this system to be developed. The focus would be to determine the "best" combination of crops, given some objective function, say minimize risk, minimize cost, and so on. The sensitivity analysis would examine the change in the optimal solution given changes to the underlying estimates. Since the data we have is based on RPUs, this system is best when applied on an RPU, regional, or national level.

Finally, the system will be enhanced through the incorporation of data from individual site locations to provide more detail in the database.

Conclusions

We have described the development of a prototype knowledge-based agricultural geographic information system for the Dominican Republic. This system incorporates an expert system shell, a database program, and a geographic information system. This integration makes the system unique in its abilities to answer questions related to both land use and crop development. Future work will expand both the area covered within the Dominican Republic, and the crops included in the database. Other enhancements to the system will include information about the economic factors of production, potential revenues, and the expertise of agricultural extension technicians. Finally, the system will be made to appear "seamless" to the user. There will be a single program that will integrate all three pieces of software and the custom programming required, and will allow the generation of a map based upon a database query or knowledge-based system consultation, rather than requiring the manual entry of commands.

This system was developed to minimize the investment required in both hardware and software. All hardware can be obtained through "mail order" type vendors, as can the expert system and database software. The most difficult aspects of this project were (and are) the knowledge acquisition for the expert system, and the data for the database. One data gathering task that we did not have to undertake was the classification of the land into RPU's. This would have been a formidable task.

This system allows Ministry officials to help evaluate, for a region, the current land use and specify which areas are being "well managed" from a land use perspective (e.g., growing bananas in an area that supports high banana yield rather than growing bananas in an arid region). This information can then be used to validate or calibrate the "potential land use function" (in an economic sense). The system will also identify other locations having the same conditions to recommend the same land use. The output of the system can be a map of the region showing where a crop is currently being produced, and areas where it should be produced, or it can be a listing of the areas. The system also provides Ministry officials the vehicle to give consistent responses as to what crops are being grown in a given area, and whether an area could support a given crop. This is critical since established procedures to do so currently do not exist, and individuals with the needed expertise are scattered throughout the Ministry. Consequently, this system will help Ministry officials establish credibility with farmers and other potential investors by giving a timely, consistent response. As they become more skilled in using the system, its benefits will grow and over time will be further enhanced by the planned additional features.

References

Barath, E., and I. Futo. 1984. A regional planning system based on artificial intelligence concepts. *Papers of the Regional Science Association*. 55: 135–154.

Bobrow, D.G., S. Mittal, and M.J. Stefik. 1986. Expert systems: perils and promise. *Communications of the ACM*. September: 880–894.

Brooks, R.A. 1983. Model-based three-dimensional interpretations of two-dimensional images." *IEEE Transactions on Pattern Analysis and Machine Intelligence.* 5: 140–150.

Burrough, P.A. 1986. *Principles of Geographic Information Systems for Land Resource Assessment*. Oxford: Oxford University Press.

Chandra, N., and W. Goran. 1986. Steps toward a knowledge-based geographical data analysis system. In *Geographic Information Systems in Government*. ed. B. Opitz. Hampton, VA: A. Deepak Publishing.

Cowen, D.J. 1988. GIS versus CAD versus DBMS: What are the differences? *Photogrammetric Engineering and Remote Sensing*. 54(11):1551–5.

Freeman, H., and J. Ahn. 1984. AUTONAP—An expert system for automatic map name placement. *Proceedings, First International Symposium on Spatial Data Handling*. Zurich, Switzerland.

Goldberg, M., M. Alvo, and G. Karam. 1984. The analysis of LANDSAT imagery using an expert system: forest applications. *Proceedings, Sixth International Symposium on Automated Cartography*. Ottawa.

Harmon, P., and D. King. 1985. *Expert Systems*. New York: John Wiley & Sons, Inc.

Laureano, E., and L. Amparo. 1985. *Agricultural zoning in the Ocoa Watershed*. Ministry of Agriculture of the Dominican Republic.

Lemmon, H. 1986. COMAX: An Expert System for Cotton Crop Management. *Science*. 233:29–33.

Miller, L.D., M. Unverferth, K. Ghormley, and M.P. Skrdla. 1989. *A Guide to MIPS: Its Features and Applications*. Lincoln, NE: MicroImages, Inc.

Morse, B. 1987. Expert interface to a geographic information system. *Proceedings, Eighth International Symposium on Automated Cartography*. Baltimore, MD.

Muller, J.C., R.D. Johnson, and L.R. Vanzella. 1986. A knowledge-based approach for developing cartographic expertise. *Proceedings, Second International Symposium on Spatial Data Handling*. Seattle, WA.

Nickerson, B.G., and H. Freeman. 1985. Development of a rule-based system for automatic map generalization. *Proceedings, Second International Symposium on Spatial Data Handling*, Seattle, WA.

Robinson, G., and M. Jackson. 1985. Expert systems in map design. *Proceedings, Seventh International Symposium on Automated Cartography*. Washington, D.C.

Robinson, V.B., A.U. Frank, and H.A. Karimi. Expert systems for geographic information systems in resource management. *AI Applications* 1(1):47–57.

Ruth, S.R., and C.K. Carlson. November 30 – December 3, 1988. Shell-based expert systems in business: a return of investment perspective. In *Proceedings of the Ninth International Conference on Information Systems.* ed. J.I. DeGross and M.H. Olson. Minneapolis, MN.

Schultink, G., B. Buckley, S. Nair, D. Brown, W. Enslin, S. Chen, J. Chen. 1987. *User's Guide to the CRIES Geographic Information System (Version 6.1).*

Smith, T., D. Peuquet, S. Menon, and P. Agarwal. KBGIS-II: A Knowledge-Based Geographical Information System. *International Journal of Geographical Information System.* 1(2):149–172.

Warkentin, M.E., P.K.R. Nair, S.R. Ruth, and K. Sprague. A Knowledge-based expert system for planning and design of agroforestry systems. Chapter 8, this book.

Waterman, D.A. (1986) *A Guide to Expert Systems.* Reading, MA: Addison-Wesley Publishing Co.

11

Diffusion of Computer Technology in Primary Health Care in the Third World

John A. Daly

Introduction

As we look to the future it seems likely that expert system technology will dramatically alter the pattern of health care in developing countries. The purpose of this chapter is to move toward a predictive model of the process of technological change involved. While the impact in primary health care is important *per se*, it is suggested that there is a need to carry out such analyses in various fields in order to identify the desirable interventions by international donors and host country governments.

Primary health care is the service offered as a patient or client first contacts a health care establishment. In the developed world, primary health care is usually offered by the physician in an office visit, but may be offered by paraprofessionals or may be offered in an emergency room or home setting. It is distinguished from secondary or tertiary care offered to the hospitalized patient. In developing countries, primary care is far more likely to be offered

This chapter is largely dependent on a previous, unpublished paper written with Antonio Ugalde, and depends heavily on insights obtained in long discussions with him. It also is quite closely related to a paper presented by the author in January 1988 at the National Academy of Sciences conference on Microcomputer Policy for Developing Countries (In Press), and benefits from the discussion at that presentation. The author also gratefully acknowledges the the research assistance of James Harold, of the Center for Development Information and Evaluation, and data provided by William Schauffler and Dora Plavetic of the Information Resources Management Office of the Agency for International Development. Thanks also to Ron Schwarz for reminding me how much this paper owes to the teaching of Jack Donoghue, and Henry Fagin.

by paraprofessionals, traditional practitioners, or even clerks in drug stores. The potential for 'skilling' the services provided by such practitioners, without greatly increasing the training they require, nor increasing the salaries they demand, has created a great interest in the application of expert systems for primary care.

When computer technology is introduced in the primary health care system, it is almost certain that there will be a variety of applications. A modular model is proposed. Hardware initially will be modest, but will as needs grow be expanded by networking and upgrading systems. Software packages similarly will be added and updated over time. Initial "gateway" applications will be followed by related and new applications. Staff increasingly will learn to transfer learning from one application to another, from one software package to another, and from one hardware system to another. Eventually, the primary care practitioner will be likely to use computer technology for a range of activities from word processing and patient scheduling, to reporting, inventory management, communications (with supervisors, colleagues, information sources, patients, diagnostic services, and support services), and even patient care. In this sense, there may be more of a role for fixed diagnostic and prescriptive protocols embodied in programs than for true expert systems that use inference engines in operation. Indeed, it seems unlikely that expert systems will by the gateway applications for the introduction of computer technology in primary health care.

In general, foreign experts appear to focus on state-of-the-art applications of technology, while workers in practical situations often focus on the simplest applications. It is not accidental that game machines and word processors have served as gateway applications of information technology in the American home, rather than higher level systems for investment planning, health care, or home education. Therefore, understanding the diffusion of expert system technology requires understanding the more general process be which information technology is likely to transform primary health care. The scale of the change under discussion is enormous. If we assume one primary care worker per 1000 population as a benchmark for future service intensity, and a population of the developing world in the range of four to five billion, there is a potential for millions of computer stations to be devoted to primary health care.

The thesis of the chapter is that the transfer of information technology to the third world health sector must be seen as a teleonomic, rather than a teleological process (Beniger 1986:41). Teleological processes are those which are goal seeking and thus seen as guided by an intelligence. Teleonomic processes also exhibit order and structure, but they are not planned. Thus evolutionary theory or socio-biology suggest that highly ordered, efficient systems may arise out of selective, programmed, or feedback processes acting on essentially random events without the intervention of a "master plan". So to, the process of the international transfer of information technology may be seen as being largely ordered and predictable, but outside the control of any one group or individual.

This is not to deny that intentionality plays a role in the process. Clearly individuals in the health system do plan and implement the acquisition of computers. Public health officials plan and implement policies to encourage the rational choice of microcomputer technology by line workers. Corporate managers plan and implement efforts to market their products. However, the overall organizational and institutional dynamic of the the international process is as far beyond the intentional control of these individual actors as is the order of the stock market beyond the control of the individual trader. The challenge is for those with effective power—such as those in responsible positions in developing country health agencies—to understand the process, and to take steps to ameliorate the process of information technology transfer that is under their responsibility.

The purpose of this chapter is to raise concerns about probable failures in the transfer of primary-care computer technology. There are a variety of different paths which potentially can be taken in the development of and use of these new health technologies. Some of these paths would improve equity (in the most important areas of well-being and survival) for the poor of poor countries, while others would accentuate inequities. Some of these would exacerbate the moral and ethical problems inherent in changing medical abilities, while others would ameliorate them. Some would enhance the power of specific players in the medical and public health system, while other paths would strengthen other players.

History suggests that the potentials are serious. In the United States, for example, there has been inflation in medical care costs, as well as national problems with health and malpractice insurance. The introduction of contraception has been related to dramatic changes in sexual morality. In mental health in this century, the introduction of hospitalization and later outpatient medication have been related to dramatic changes in the roles and values associated with mental disease. Similarly, "counterfactual inquiry" suggests that the great human suffering that followed in the wake of the industrial revolution was not intrinsic to the technological change, but was a function of the social and political choices made (Perrow 1986:272–278).

In like manner, experience with information technology suggests that some decisions taken early in the process of institutionalization have serious repercussions, and are essentially irreversible. Thus the decision to emphasize public broadcasting in radio or television, as was done in Britain, or private broadcasting, as was done in the U.S., has great impact on the nature of the programming and its public impact. Moreover, the enterprises that resulted from the decisions formed powerful lobbies for their own continuation.

FIGURE 11.1 Conceptual Framework

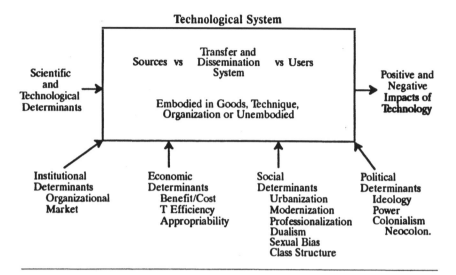

Similarly, the free market decisions that resulted in emphasis of high cost computer applications in health, such as CAT scanners, resulted in the creation of a lobby for financing of services embodying the devices. If such decisions can be made more thoughtfully in developing countries, perhaps the final impact of the introduction of computers will be more beneficial.

The paper is organized according to the scheme shown in Figure 11.1. The technological system, involving the sources, disseminators and users of technology is the central focus. In health-related information technology, both the health sector and the informatics sectors must be considered. Central to the discussion is the way the embodiment of the technology affects the transfer process. In seeking to understand the possible positive and negative impacts which may result from present and potentially available technology, the discussion focuses on scientific and technological determinants, institutional determinants, economic determinants, social determinants, and political determinants.

Technology Transfer

For simplicity, we will consider the health technology system, to consist of three main components: the technology development system; the technology transfer and diffusion system; and the technology dissemination system. Each of the three systems is itself a loosely interacting set of interrelated subsystems. This conceptualizaton is based partially on that developed by the U.S. Office of Technology Assessment (1976) and the studies carried out in support of the President's Biomedical Research Panel (U.S. DHEW 1976). It also draws heavily on more general studies of technology innovation (Tornatzky et al. 1983) and diffusion (Rogers 1983). The health technology system may be considered in part a subsystem of the health sector discussed above, but it also includes sources and intermediaries outside of the health sector. One important source for health-information technology is the information technology sector.

If technology is knowledge, it can be transferred in the form of information contained in some media—books, radio or television programs, instructional cassettes, person to person training or professional exchanges, journals, and so forth. Such media embody technology or technological knowledge (Price 1986). Information technology is particularly important in its potential for increasing the international availability of this unembodied technology. Satellite technology potentially allows health practitioners in even the poorest countries to access data and data bases in the most advanced countries. Moreover, computer networks are making interactive two way communication almost as low cost as one-way mass media communication for some purposes (Pool 1984).

In similar fashion, goods embody technological knowledge.[1] For example, a new pharmaceutical product may be considered to embody the knowledge of biochemistry, pharmacology, pathology, microbiology and physiology that went into its discovery, development and testing. The patient and the practitioner do not need to internalize this knowledge in order to utilize the drug in which it is embodied. An antibiotic works as well if the patient does not know from which organism the genes for its production were first discovered, in which organism it is produced, or how it works. Similarly, one might describe iodization of salt for the prevention of goiter, or vitamin A fortification of oil as embodiment of health technology in food goods.

The study of technological innovation and diffusion has been particularly strong in the pharmaceutical area (Mansfield 1971; Jadlow 1976). Such studies typically are within the discipline of economics, and explore hypotheses on the effect of market strength on innovation, and innovation on market position. Similarly, there are some classical results on the diffusion of knowledge about new pharmaceuticals among physicians (Coleman, Katz, and Manzel 1966, 1957). These have contributed importantly to the understanding of the social factors influencing the diffusion of innovations within a community. In terms of knowledge embodied in goods, computers are especially interesting

---

[1] It is usual in the health field to distinguish between that technology embodied in medical devices—surgical instruments, x-ray equipment, CAT scanners, stethoscopes—and that embodied in supplies such as pharmaceuticals, diagnostic reagents, fluids, etc. (National Academy of Sciences 1979; Lukas 1978). Medical devices are usually seen as capital expenditures, to be operated by relatively highly trained health service personnel. There are also specialized studies of technology transfer in the water and sanitation field (Crain et al. 1969; Beverly 1984).

because artificial intelligence techniques allow knowledge to be embodied symbolically in machines that can adapt and modify that technological knowledge, and can proactively interact with humans to interchange knowledge and information.

Technology also is embodied in processes. A great deal of scientific research and technological development has gone into the development of fermentation processes that produce antibiotics with high purity and at low cost. In large part this technology can be embodied in the equipment and rules for manufacture, so that the factory workers can control effectively the manufacturing process without full mastery of the many disciplines that went into the design of the process. The international transfer of processes for the manufacture of pharmaceuticals is itself a major field of study (c.f Cilingroglu 1975; National Academy of Sciences 1973; Botero 1972; McCraine 1976; Developing World Industry and Technology, Inc. 1979; Organization of American States 1982; Office of Technology Assessment 1986).

Perhaps most importantly, technology can be embodied in people. Typically this is described as "technique". Thus the diagnostic technique of the internist, the operating technique of the surgeon, the laboratory technique of the medical technician, the hygiene techniques of the sanitarian, the entomological techniques of the malariologist, the delivery technique of the traditional midwife, and the techniques of the mother in the diagnosis of illness of the child may all be classified as human embodied health technology. The training of health professionals and health education both represent investments in human capital, involving the increase and improvement in health technology embodied in people. There is also a major literature dealing with the transfer of technology through participant training (Steele and Oesterling 1984).

Finally, technology may be embodied in organizations through organizational norms and standards, procedures, and structures and result in the production of goods and services that embody technological knowledge beyond that of the members of the organization. Thus health service organizations that implement services developed in other organizations, or that implement the recommendations of outside evaluators or advisors, may be considered to embody technological knowledge in the organization. There is a relatively limited literature on the transfer of such technology (U.S. DHEW 1971; Rothman 1980).

Sources of Technology

If we accept the definition of health technology as knowledge of health practices, some health technology comes out of distant history—the superstitious beliefs in hot and cold as sources of health and disease from Greek philosophers, use of quinine for malaria and willow bark (leading to the discovery of aspirin) for fever and pain from South American and European herbalists. Smallpox inoculation apparently originated in Asia, was in the early 1700's introduced to England through Turkey (through the auspices of Lady Montague, wife of the English Ambassador), and in the early 1800's was replaced by vaccination as a result of the studies of Edward Jenner, an English physician-naturalist (Hopkins 1983:46–81).

Certainly, however, the rapid development of health technology in the last century is geometrically increasing due to the use of modern scientific and technological institutions for the development of health technology. The development of the university based scientific medical school is surely a major element. First in Europe, in the United States in the beginning of the 20th century, and increasingly in developing countries medical schools were seen as research as well as training centers.

In the United States, dating from about the Flexner report (1910), private medical schools were forced into closure by increasingly stringent accreditation standards, and university medical schools increasingly controlled the training of physicians (Starr 1982:112–123). In part this could be seen as a social process that, by defining the creation of new technology as integral to the training of doctors, forced the funders of medical education to pay for the development of the technology to be embodied in human resources. In the 1950's and 60's, the U. S. National Institutes of Health tapped an enormous vein of federal subsidies for biomedical research and channeled it to schools of medicine. The non-clinical, basic research establishment of the schools grew to the point that they threatened the control of the medical schools, and threatened to change the principle purpose of the schools (Starr 1982).

Interestingly, no similar development of other schools of health providers seems to have produced comparable research capacity. For example, as compared with 126 medical schools in the United States in 1980 (Starr 1982), The number of schools of public health grew from one in 1916, to 12 in 1965, to 18 by 1975 (Bowers 1975). Similarly, nursing schools and schools for other allied health personnel do not seem to have a research focus comparable to the medical schools.

A second major, modern source of health technology is the corporate research laboratory. Pharmaceutical manufacturers and medical device manufacturers internalize much of the research and development on their products and processes in their own corporate laboratories, as well as contracting (typically more generic) research to universities. To some degree private research laboratories also contract research services to these industries.

Historically the Rockefeller Laboratories were prototypical of non-profit laboratories heavily involved in medical research. In the last half-century in the United States, however, government laboratories have grown enormously. Prototypical now would be the intramural laboratories of the National Institutes of Health, but other U.S. agencies such as the Department of Defense and the Veterans Administration also have research laboratories. With regard to the development of technology for developing countries, there are also a few international laboratories, such as the Diarrheal Research Laboratory in Dacca, supported by a variety of governmental and donor agencies.

Health technology need not always come from health laboratories. Agricultural laboratories, in their work on zoonoses, nutrition, and hygiene have developed important technologies for human health. Historically, in the 19th century in the United States, for example, there was much greater funding for livestock than for human health research. As Shyrock puts it:

> It seems ironic that federal aid was extended in this period to studies of the health of farm animals, while almost no funds were available for direct work on the diseases of man. For the most part the public saw no possibilities in the latter program, partly because of the nature of medical science prior to 1885 and partly because human welfare brought no direct financial return. Hogs did. (Shyrock 1980:44)

Even today international agricultural laboratories, such as ICIPE in Africa, are better funded and more powerful than international health laboratories, and they, too, do health related research of importance to the tropics.

In terms of the specific orientation toward microcomputers and microelectronics, a mixed pattern of laboratories exists, and new forms appear to be evolving specific to these fields. Microelectronics hardware research is clearly focused in industrial laboratories, but software R&D is found in all locations. An interesting development is the creation of private centers for generic research by groups of companies (MCC in the U.S.), and/or international programs for such research by groups of countries (e.g. ESPRIT in Europe). Increasingly the line between the universities' traditionally non-profit research and the corporate research is being challenged, both by faculty with a foot in both worlds and by new forms of corporate support for university based laboratories.

Users of Technology

We may consider several classes of users of health technology: the individual; the traditional practitioner; the modern practitioner; novel practitioner categories; organizations. The individual may treat himself or herself, or comparably, one member of the family may treat another. In the North the access of individuals to health technologies has been increasingly modulated by professional gate-keepers. Access to pharmaceuticals is limited by prescription laws, placing control in the hands of pharmacists, physicians, and in some cases specialized practitioners such as dentists. Similarly, access to information for diagnostic or other "practitioner embodied" techniques is usually written in specialized journals and terminology that limits lay access, and encourages practitioner consultation. In less-developed countries, individual access to some medical technology is less limited, and there may be a far greater tradition of self-medication.

Traditional medical practitioners, while almost totally supplanted by physician-based modern systems in the North, are still extremely important in the developing world. It has been estimated that 60 to 80 percent of babies born in the developing world are delivered by midwives (Population Reports 1980,1). Further, there are many differing categories of such practitioners, including religious practitioners, bone setters, curers, specialists in acupuncture, and herbalists. It would be incorrect to assume that these practitioners are similar from culture to culture, that they share a common "primitive" nature, or that the technology is unchanging. For example, Ayurvedic medicine, practiced in India and nearby countries, is an ancient and classical system, based on written texts, and taught to practitioners in centralized schools (Pillsbury 1978; Fraser 1979). Homeopathic medicine, in contrast, originated with the theories of Samuel Hahnemann (1755–1843) and by 1890 involved specific approaches to the testing of new drugs (on healthy subjects), had expanded to thousands of practitioners and 16 medical schools in the U.S., and eventually was integrated by technological convergence into modern or "allopathic" medicine (Starr 1982).

Modern practitioners include physicians, nurses, nursing aides, dentists, physician specialists. It might be useful also to include some related professional categories, such as civil engineers involved in the design of community water supplies and sewerage systems, veterinarians involved in sanitary inspection, electro-mechanical engineers involved in the maintenance of medical devices, and hospital and public health administrators.

In developing countries, there has been a major effort to create new categories of health workers in the public health system, generically called paraprofessionals. These may be physician extenders, working under relatively direct control of physicians, health promoters with relatively little formal training, community sanitarians, and so forth. There are also, however, new groups such as injectionists—who without standardized training or licensing provide injections of antibiotics, vitamins, or other substances—who function on a fee-for-service basis and who can find a market for relatively modern technology in the absence of more professional infrastructure.

Finally, organizations ranging in size from individual hospitals, to small decentralized health institutions, to ministries of health may also be considered as users of health technology. In such cases as the purchase of a CAT scanner or a Cobalt bomb radiation therapy machine, only a relatively large organization has the capital and demand to acquire the technology. The organization may make the acquisition decision in ways that are not under the control or even full understanding of any individual. Similarly, when a health delivery system including a number of hospital and outpatient facilities specifies a drug formulary and centralizes the purchase and distribution of pharmaceuticals, it may be most appropriate to treat the organization as the user.

The Technology Transfer and Diffusion System: In terms of the definitions above, the technology transfer and diffusion system is necessarily large and expensive. For a health technology to work it is usually necessary to consider a body of technology embodied in several different places. For even as simple a product as an antibiotic to be useful, it is necessary to have the right drug in efficacious form (i.e. the correct antibiotic, which is still active), and the diagnostic knowledge (usually embodied in a gatekeeper) of the signs and symptoms calling for its use, as well as the knowledge (embodied in the family) of the dosage and timing of applications. To get such knowledge from hundreds of manufacturers to millions of practitioners, and to billions of patients is a very costly undertaking.

Commercial markets are an important technology transfer system. They are the principal vehicles for the transfer of goods embodied technologies. Thus the international transfer of pharmaceuticals and medical devices is handled through international market mechanisms, and the internal distribution is handled through systems of wholesale and retail markets—prototypically the network of drug stores. However, markets also can be used for technology embodied in people and organizations. Thus individual consultants and consulting organizations bid for technical assistance contracts for construction of water and sanitation works.

It should be noted that while specific markets have been institutionalized in most countries for goods embodying health technologies—pharmaceuticals and health technology—other market institutions can expand to disseminate such technology. Thus contraceptive social marketing has experimented explicitly with use of commercial food distribution channels (Population Reports 1985). Similarly, AT International has experimented with commercial metal working workshops as a vehicle for the transfer of wheelchair technology based on sports chair designs and novel materials (J.G. Smith, pers. com.). The Water and Sanitation for Health (WASH) project team and Georgia Tech. have experimented with commercial foundries for the transfer of new sanitation technologies (J. Beverly, pers. com.).

Educational and professional channels are important in the transfer of technology embodied in people. Enormous numbers of health professionals receive general and specific technical training in foreign countries; donor agencies provide technical assistance teams to developing countries; scientific and professional exchanges are common; international professional meetings, professional journals, and similar mechanisms are patronized by the professional societies; medical texts are internationally distributed. The importance of such channels can be illustrated by the fact that the Pan American Health Organization may well be the world leader in developing computer technology for the translation between English and Spanish. Comparable channels within countries are obvious.

In the training of traditional practitioners, apprenticeship historically was the rule. For example a midwife or herbalist worked for a practitioner until he or she obtained the necessary skills to go successfully into practice for himself or herself. Until the 18th century, training of surgeons was through such processes under the control of guilds even in Europe. Increasingly, however, formal educational procedures have been used to train or upgrade the skills of traditional practitioners. Thus, for example, Egyptian Dayas have been trained by the government since at least the 1940's (Pillsbury 1978). More generally, a survey of literature on traditional midwives found 34 documented programs or experiments in the training of these practitioners (Population Reports 1980). It seems clear that formal training in classroom and clinical environments is a superior channel for rapidly embodying technological changes in health workers.

In providing technological information to families and individuals in developing countries, there is considerable potential for health education services. This may be formal in the schools, or informal through health education services of the health agencies. It is also possible for simple techniques, such as contraception and oral rehydration therapy, to use mass media, drawing upon the communications technologies developed for advertising in the North (Population Reports 1985).

Finally, organizations provide channels internal to those organizations for the transfer and dissemination of technology. The importance of such organizational channels has always been recognized for multinational corporations involved in the sale of pharmaceuticals and medical devices. Similarly, the utility of international organizations for the communication of certain quarantine and standardization technology has been recognized for nearly 100 years, as attested by the history of the Pan American Sanitary Bureau, the Health Organization of the League of Nations, and the World Health Organization. In the last decade, however, health care provider organizations have begun to take a significant role in the international transfer of technology. Within developing countries, ministries of health, social security agencies and similar organizations play a major role.

Again, the technologies built around microcomputers and micro-electronics to some degree will share the channels used traditionally, but may be expected to require and to generate new technology transfer mechanisms. A great deal of microelectronic and microcomputer technology will be embodied in products, and will be disseminated through the existing product related channels. Information technology is in the lead, however, in developing new channels and there already is great interest in the opportunities it provides for electronic networking for the transfer of technology. On the other hand, in the developed world enormous networks of computer users have evolved who exchange software and technological information without formal organization, market mechanisms, or subsidies.

In summary, the embodiment of the technology and its intended user greatly influence the systems for its generation and the choice of dissemination channels. Prototypically, as shown in Figure 11.2, technology embodied in goods is developed in commercial laboratories and disseminated through market channels. Equally prototypically, technology embodied in people is developed in governmentally or charitably subsidized laboratories, and disseminated through educational and professional channels.

Scientific and Technological Determinants

In the introduction of computer technology for primary health care, the first determinants to be considered are scientific and technological. The current state-of-the-art is considered, as are potential advances. The potential includes the prospects for low-cost, rugged hardware and software in native languages of developing countries. In the case of expert systems, we discuss the potential for development of specific expert systems for tropical diseases.

Obviously if an application is not technically feasible, it will not occur. However, just because an application is feasible does not mean that it will occur. To underline the obvious, the status of biomedical sciences dramatically affects the potential in the use of expert systems in primary health care. As current understanding of the immune system and parasitology increases, new

FIGURE 11.2 Transfer of Technology in Primary Health Care System

0087D					
	Embodiment of Technology				
	Media	*Goods*	*Process*	*Technique*	*Organization*
Source	University Laboratory Professional Assoc. Foundation	Pharmaceutical Industry Med. Equip. Industry Computer Industry	Pharmaceutical Industry	University Univ. Hosp.	Health Ser. Res. Agency Government Health Ser. Industry
Transfer and Diffusion System	Professional Assoc. Market W. H. O. Media	Market Organizations	Market	Professional Assoc. W. H. O. Donors Health Ser. Orgs.	W. H. O. Health Ser. M.N.C.'s Professional Assoc.
Gatekeeper		Physician Pharmacist		Physician	
User	Individual Family Practitioner	Individual Family Practitioner	Enterprise	Practitioner Individual Family	Health Ser. Res. Agency Government
Regulation	F.C.C.	F.D.A.	F.D.A.	Licensing	

diagnostics are developed for tropical disease identification, new vaccines for their prevention, and new understanding for their treatment. These new developments dramatically influence the content of expert systems for tropical medicine. New scientific knowledge and technology for a disease like malaria may allow a malaria module to be written for a paramedical expert system, where early technology would have required a specialized professional.

The case of surgery is a useful historical example of the technological requirements for the development of health technology (Cartwright 1967). Anatomy and pathology provide the basic scientific underpinning for surgical intervention. Still through the 19th century, surgery was a rare procedure, practiced by a lower class of practitioners, with grave risks to the patient. Anesthesia was invented in the 1840's; antisepsis was developed in the 1860's; and diagnostic X-rays discovered in the 1890's. Only then did it become possible to develop the instruments and techniques characterizing modern surgery. In like manner, Babbage's development of computers in the 19th century failed because the available mechanical technology could not produce an economically efficient computer. His work was rediscovered after the reinvention of the computer, in a more conducive technological environment based on electronics.

In many cases, modern technology offers the first potential for what Lewis Thomas has called "high technology": "the decisive, conclusively effective measures that can be aimed directly at the underlying cause of a disease, so that it can be terminated, reversed, or best of all prevented outright." (DHEW 1976:4) Certainly the enthusiasts for new biotechnological vaccines promote them in this vein. However, it also has the potential for the delivery of what Thomas calls "halfway technology": expensive, complex techniques for the amelioration of common diseases. For example, it seems at least theoretically possible that new vaccines against protozoa will be only partially effective, and that the populations of these organisms will evolve resistance, requiring the use of the vaccines in a complex system of preventive and curative measures to effect overall control. Similarly, it appears conceptually possible that a complex technology could be developed tailoring individual drugs to individual cancers through the combination of monoclonal antibodies and toxins. This could occur without real understanding of the mechanisms involved, and involve great expense in the production of the therapeutic agent and in the control of the impact of the therapy on the patient.

The Specific Determinants Related to Computer Technology

The driving force in the growth of computer technology is the trend, maintained for four decades, of increasing computer hardware power at decreasing cost. This steady decline in the cost of hardware has created a huge and growing market for computer software. New classes of software products, popularized in the last decade, such as spreadsheets, word processors, data base managers, and communications packages are sold in the hundreds of thousands and sometimes millions of copies. Even ignoring pirating, it becomes theoretically possible to distribute the cost of software production, distribution and maintenance over very large numbers of users, and thus make per-unit software costs relatively small. Nonetheless, software development and maintenance costs now greatly exceed hardware costs, reversing the situation that dominated in the 1960's (Schware 1987:1253).

The dissemination of computer technology may be expected to follow a familiar pattern. Within a homogeneous group of users, there will be some early acceptors, and some late (Kraemar and King 1984). Moreover, the poorest countries are only now adopting computer technologies that were adopted in the United States in the 1960's and 1970's (Kraemar and Perry 1979).

The pattern of overseas installation of computers in the offices of my own agency illustrates the likely pattern within a large organization (or within a specific market). Minicomputers were introduced in essentially all overseas offices in the early 1980's, while the numbers of workstations and stand-alone microcomputers has continued to increase.

Applications will differ in acceptance. Some will be almost universally accepted, while others will be disseminated among only some or a few users. However, as time continues, the total number of applications per user organization will increase, and the number of application-workstation pairs will increase even more rapidly.

The most common use of computers in primary health in developing countries appears to be in managerial and administrative functions (National Academy of Sciences 1986, Ch. 4). It seems likely that these applications will become much more wide-spread as computers become cheaper, software more available, and demonstration efforts more successful and visible.

Computer technology is revolutionizing biomedical instrumentation (Eden 1986). Chips built into many sensors digitize and display data. New materials and instrumentation technology potentially allow the development of a variety of new medical sensors. Processors transform data into forms easier to interpret. In extreme cases—such as computer-aided tomography and magnetic resonance imaging—the computational power now available has made possible the introduction of powerful, but expensive, medical devices. One may predict that the potential exists for an enormous range of new instrumentation, including new, low-cost instruments such as the PATH electronic scale, and new and even more expensive imaging technology.

In the area of robotics, already there are automated clinical laboratory devices. It seems likely that as this technology improves generally, there will be additional potential for the automation of smaller scale tasks and more advanced tests. In other sectors there is continuing interest in developing self-controlled mobile devices, which may eventually have health applications.

The earliest of these probably will be in the processing of hazardous materials and work in hazardous environments. Still, Japanese researchers are already experimenting with a robot that assists the nurse in heavy tasks associated with patient care (Hashino 1986).

Microcomputer technology is making possible new dimensions in epidemiological and health services research. Thus, a recent health status survey done in Rangoon Burma was in analysis within four days of the beginning of field work, and the final report was submitted to the Director of Public Health on the tenth day (Frerichs 1987).

In the future, it has been predicted that computer applications will involve more natural language capabilities. Thus the Pan American Health Organization, with Agency for International Development support, has already developed a functioning system for the translation between Spanish and English. Systems exist which respond verbally to verbal questions received by telephone (Daly 1986). Similarly, AID has funded the development of an expert system which helps paraprofessionals recommend triage and treatment for patients with eye problems. (Kastner et al. 1984) It is rumored that the PRC has developed more than 100 such expert systems for different aspects of paramedical practice, and is equipping paramedics with Chinese made microprocessors in the field. On the other hand, very elaborate expert systems are being developed in the U.S. to aid physician-specialists in their work.

Thus the potential exists for developments in preventive and curative health services; in administrative, clinical, laboratory and scientific applications; at low and at very high cost; relevant to the needs of patients with acute or chronic disease. The potential extends to all segments of the health sector. An optimistic scenario for the impact of microelectronics in health suggests:

- Improved data management will improve medical record keeping and medical research; medical knowledge will increase dramatically.

- Improved sensors and data handling will improve medical diagnosis, making medical attention more timely and therefore more efficacious.

- Scheduling and administration of health care delivery will be improved, as will laboratory services. The time saved by providers will be reflected in more personalized and professional care.

- Computers will allow centralization of medical records, greatly improving emergency medical care, and allowing patients to be followed in order to schedule preventive and health maintenance services optimally.

- Life expectancy will increase, both as a result of the above innovations, and as a result of uses of information technology in other sectors which reduce health risks to the population (Textor et al. 1983).

As any powerful technology, microelectronics technology offers dangers. These of course include the dangers of willful misuse. There are also dangers, however, of accidental side-effects of the use of the technology. For example, one could easily imagine the dehumanization of medical services, if automation were introduced without regard to the historical and cultural functions of the practitioner-patient interaction. Similarly, as the body of knowledge embodied in computer devices becomes greater, and more opaque to the practitioner and user, the potential for error and use of outmoded information increases. Thus Textor's pessimistic scenario includes:

- Patients being treated by machine—inhuman, impersonal, demeaning, and ineffective. The human factor largely eliminated in assembly line medicine.

- Large corporations use central data banks to screen the population in order to discriminate on medical grounds in employment.

- Increased pollution and personal risk from increased industrialization.

- Increased mental problems of *anomie* from the growing isolation of workers whose contact is increasingly mediated by machines.

- Increased work stress in an ever more automated and impersonal world.

- Widespread disorientation and anxiety caused by information overload in the "global village" (Textor et al. 1983).

Institutional Determinants

Primary health care is organized into several subsystems in most developing countries, and each subsystem has its own vehicles for technology transfer and diffusion. Thus, one may predict the rate of introduction of computer technology in a subsystem, and accept that rate or reject it as inappropriate. If one believes that a faster or slower rate of introduction, or a higher or lower quality of introduced technology is appropriate, the intervention required will depend on the specific institutional system involved. Three examples are given to illustrate the form of the analysis.

Large Scale Organization

In all developing countries there are large scale formal organizations that provide primary health care. These include public health services, operated by national governments (Ministries of Health), Social Security Institutes, Military Health Services, and in some cases prepaid medical service organizations. In some cases, a single organization will provide service varieties of primary health care, as when a public health service offers physician-based medical services in a system of health clinics and dispensaries, and paramedical-based services through home and community visits. In such organizational settings the rate of introduction of computer technology will depend on the rate of capital formation—poorly funded services will lag.

There is, of course, a significant literature on the introduction of new technology in large scale organizations (Zaltman 1973; Heise 1981; Brodman 1988). If one wishes to increase the rate of innovation, one can: (a) define formal policies to that effect; (b) create permanent or *ad hoc* units to promote the innovation; (c) evaluate units, supervisors and workers on the degree to which they have encouraged appropriate innovation and provide economic incentives for the successful innovators; (d) increase the rate of investment in hardware, software, training, and technical assistance for technological innovation; (e) hire new staff with appropriate training and skills, or acquire new units, such as computer departments, to add technological strength to the organization. Typically, specific organizational processes will be created for provision of supplies, maintenance of equipment, continuing training for user personnel, and for the development and maintenance of new software applications.

In short, a modern medical care organization internalizes a variety of mechanisms for technology acquisition and dissemination. The organization is not, however, simply neutral as regards technology acquisition by its various parts. Simply because the public health service in a country may operate multiple services—such as physician based primary health care, health promoter based home visits, and midwife based prenatal care services—all three probably will not have equal access to computer technology. Within the organization, there may be significant impacts of social, economic and political factors as described below. Similarly, there are great differences between different health service organizations in the same country, and one can expect that wealthier organizations, employing higher status professionals, with higher status clientele, will acquire computer technology more rapidly than their poorer, lower-status competitors.

Organizational theory has tended to promote garbage-can theories of management, suggesting that management decisions are not normatively rational, but are rather the process of groups satisficing the various constraints and objectives which they face. In the case of developing countries, it often appears that a health service organization is more permeated by objectives to provide employment, visibly demonstrate government presence, and serve the personal needs of employees than in the U.S. Thus, the lack of "technological efficiency" in introducing new techniques to improve the cost effectiveness of health services in part may be due to the fact that this is only a partial objective of the organization.

Fee-for-Service Primary Health Service Practice

In many developing countries almost all physicians are in private practice, at least a part of the day, even though their primary employment may be in organized medicine. If one wishes to promote the introduction of computers into this private practice, a number of policy instruments are possible. These tend to be in the private sector (although, as physicians in the public health organizations become experienced in the use of computers in their work, they will tend to transfer the technology to private practice). Thus, one might change tariff duties for the import of hardware and software for private medical practice, subsidize the acquisition of hardware and software through tax subsidies or special loan instruments, or organize mass acquisitions of hardware and software through medical associations to secure low prices.

Similarly, one might seek to stimulate training of physicians in the use of new systems. This could be done through the continuing education services of medical schools or medical associations, or could be done through incentives for the pharmaceutical distribution agencies, which have heavily invested in systems to reach physicians with information, including networks of hundreds of thousands of pharmaceutical detail men. The licensing power of the state could be used to encourage physicians to obtain the training, and to standardize the software and expert systems that were introduced.

Obviously, there could be subsidies for sources of this technology as well as for its users, combined with subsidies for intermediaries. Thus, it would be possible for the government to subsidize research and development to produce expert systems for physicians in the country. Similarly, government could encourage the creation of private sector enterpriss that sell computer services, software and software services, and hardware and hardware maintenance to the physician in private practice. Such encouragement could be in the form of legal authorization, tax incentives, loans, equity investment, or lowering of disincentives to co-ventures from abroad in these areas.

Note that, while not likely to occur without outside intervention, many of the same approaches could be applied to encouraging the use of computers by others in private practice, such as midwives or traditional Chinese or Ayurvedic practitioners. However, medical practice has evolved systems of continuing education and technology transfer through professional meetings and licensing that are most amenable to technological innovation. More traditional groups are not so blessed.

Pharmaceutical Dispensaries

The network of sales points for pharmaceuticals in developing countries can include itinerant vendors working rural markets, and local general goods stores as well as pharmacies *per se*, and pharmaceutical dispensaries in health facilities. In many countries there exists a private sector system of pharmacies, often concentrated in urban areas, and staffed by reasonably well-trained and experienced personnel, if not by professional pharmacists. These pharmacies inescapably provide customers with advice as to the appropriate medication for the presenting symptoms, and all too frequently provide medical care including prescriptions and injections of prescribed medications. If one wished to encourage the use of computers in these pharmacies, and the introduction of expert systems for the advice of patients, it would be quite possible to do so.

Strong networks link these pharmacies with the pharmaceutical distribution network, and that network could be used to provide technological inputs, including hardware, software, and training. Similarly, one would expect to find a trade association of retail pharmacies, which could provide linkages for training or for standardization. A trade association could also serve to promote low cost acquisition of the technology through the choice of the technology and through monopolistic purchasing for its members.

Many of the policy instruments—such as tax incentives and the creation of service institutions, discussed for the fee-for-service physician network—would be equally applicable to encouraging use in private pharmacies. On the other hand, some instruments would be specific to the subsector. Thus, the development of appropriate computer systems for pharmacy stock control and management could be subsidized, as could the development of simple client record and billing systems and drug interaction systems. Such actions to encourage the dissemination of computers in pharmacies for other purposes would facilitate the introduction of computerized primary health care advisor applications in the pharmacy. Similarly, legislation to reduce the legal liability of the pharmacy for malfunctions of such systems would remove a potential disincentive.

Environmental Theories

Some theorists (Hirsch 1975) have suggested that the technological environment is part of the general institutional environment. As technology changes, leaders and organizations seek to adapt to the new circumstances. Thus one can suggest that the proliferation of software and knowledge engineering firms and of biotechnology firms is a predictable industrial response to the uncertainties involved in these new technologies. Multinational corporations can allow the many small entrepreneurial firms to maximize innovation—and incidentally to bear the risks of failure at the R&D stage—while maintaining control of manufacturing and distribution channels. One would predict that as the technologies mature, large MNCs again will extend control over the technology generation. Similarly, the ability of a new set of companies to capture large parts of the microcomputer market can be attributed to the predictably slow response of the large companies to entering into a new market created by the technological innovation of the microchip; the relatively rapid capability of IBM to capture a part of this market as it matured would be equally predictable. Finally, the rapid series of mergers between seed and chemical companies, and between computer and communications companies is also predictable within this school of organizational studies.

If this is true, then one might seek to encourage a similar institutional pattern in the area of primary health care applications of computers. Thus

small groups might be formed to do the innovative, high-risk work of developing new applications. Larger organizations, such as health service agencies,
pharmaceutical distributors, and computer hardware firms could serve as the
major dissemination channels, allowing them to appropriate benefits from the
end-user commensurate with the cost and risk in the dissemination.

Structural/Functional Analysis

The same institutional structures can perform different functions in
different countries or at different times. Thus in the U. S. over the past 40
years, massive government funding of medical research in the universities has,
especially in the top research schools, biased the function toward knowledge
and technology creation. In developing countries, the university remains most
heavily devoted to the training function, and secondarily to public service, with
research little emphasized. However, in specific countries universities have
taken on other functions such as the fostering of a political opposition, or the
co-optation of a new middle class to support the government. In general, the
result of these two tendencies is to increase the intellectual distance between
U. S. and LDC universities, and to make the LDC university a far less technologically inventive institution.

Sectorial and technological institutions are less functionally distinct from
political institutions in LDC's than in developed countries. Social power in the
health sector in developing countries is more fully political than in developed,
and key decisions such as the location of a new hospital or medical school, or
the enrollment of that medical school are frequently made for political and not
"technical" reasons. Consequently there is relatively less demand and value
attached to technical planning. Traditions tend to be authoritarian rather than
democratic, and the public defense of the rationale for a decision is less
detailed and comprehensive. This suggests that the role of computers in decision making will be less valued.

Economic Determinants

The fundamental economic principle in the development and dissemination of health technology is that the benefits from the new technology should
be consonant with the costs involved. For any given innovation it is likely that
the rate of expenditure will start low, as researchers search for an idea, will
increase as the the innovation comes to the point of clinical or commercial
application and as it is field tested, and increase again as it is diffused to large
numbers of users. For any given innovation, the risk of failure is high.
However, major laboratories will have many innovations in development at
the same time and amortize the losses of the unsuccessful attempts over the
returns from the successes. Estimates of returns from biomedical R&D range
from very high to quite modest.

More generally it is believed that the average rate of return from invest-ment in technology depends on the circumstances at the time. In part these circumstances are policy dependent, as in the case of R&D investment credits or patent protection. In part, however, they appear to depend on uncon-trollable circumstances. Thus it is generally assumed that there are stages of relative maturity in any industry, during which the potentials in existing scien-tific information have been exhausted and marginal rates of return on R&D rapidly decline. If the hypothesis is correct that microbiology and microelectronics are new, revolutionary technologies, then marginal rates of return for health R&D based on these technologies should be high in the fore-seeable future.

In theory it is possible to rank order information technology applications in the health sector in terms of benefit to cost ratio. In general, one would expect to initiate the development of computer capacity by introducing the most cost-beneficial applications first. Thus one might seek to introduce computers first to rationalize the allocation of resources for large organiza-tions, to increase the efficiency of the scarcest resources, and to make the most beneficial increases in quality of service. However, there are considerable externalities in the process of introducing information technology, and consid-erable uncertainty unavoidable in the decision process. Therefore, initial innovations may also be those involving low cost, low risk, high visibility, or strong support.

The "efficiency" of a market is defined in terms of the instantaneous modification of prices to reflect new information. We define "technological efficiency" to be the instantaneous substitution of the most appropriate tech-nique given the current demands and factor prices, involving the most current scientific information. Clearly the shorter the lag in the application of a health technology, the more likely the technological investment will be recouped in appropriate benefits. Historical evidence suggests that the lag between discovery and application in the health field is variable, with many innovations applied within a year, most within ten, and some requiring much longer (Comroe 1976). It is difficult to generalize about the "ideal" lag which allows for adequate clinical testing and does not force the introduction of an overex-pensive technique. Nonetheless it appears likely that lags are longer when a technology must be transferred between countries. In one study it appeared that technologies were transferred more rapidly to Asia than to Latin America, and less rapidly to Africa (Piachaud 1979).

"Appropriability" is the degree to which the benefits accruing from the use of a technique can be captured by a party. In general it is necessary for the developers and disseminators of a technological innovation to be able to appropriate adequate portions of the benefits to continue to function. However, "information is a durable good, in that present resources must be devoted to its creation and its existence results in a stream of future benefits.

Information is also a public good, in that once it is created its use by second parties does not preclude its continued use by the party who discovered it. However, use by second parties does reduce the private return on information created by the first party. This is the "appropriability problem" (Magee 1977).

In general, when technology can be embodied in goods the appropriability problem is lessened. A drug may be used only once, a medical device in use by one person precludes its simultaneous use by others. Thus the developers and disseminators of goods-embodied technologies can appropriate a portion of the benefits they bring consumers through the sale of the goods. In the west commercial firms dominate and market processes are the key transfer mechanisms for such health technologies. Where market processes are imperfect government can subsidize (as is often done in R&D for "orphan drugs") or regulate.

Technology embodied in human resources is a more difficult problem. Licensing of physicians, which is a relatively new social invention, allows the appropriation of relatively high training costs through the sale of services. Similarly, a social invention of the early 20th century was the use of medical school fees to subsidize the development and dissemination of medical techniques in research universities. It is suggested that these market oriented techniques are relatively less successful in the case of practitioner-embodied than goods-embodied technology.

The development of charitable foundations about the beginning of the 20th century provided a profound stimulus to health R&D in resolving a portion of the appropriability problem. Philanthropists like Rockefeller and Carnegie devoted portions of their immense fortunes to endowments to subsidize medical research; much of the research was of direct or indirect benefit to developing countries. However, with exceptions such as the Carvajal Foundation in Colombia, few developing country foundations have followed this pattern.

More recently governments have used tax powers to provide subsidies for health R&D. The majority have been direct subsidies, but tax reductions for charitable contributions and corporate R&D investments have also been significant. In the international arena the combination of foundations and governmentally supported donor agencies have become important, both in the transfer of technology through organizations such as WHO and in the development of technology in programs like the Tropical Disease Research Program managed by WHO and the World Bank.

It is interesting to note that major advances in health technology have often occurred in institutions that had alternative ways to appropriate benefits. Thus research on livestock health—funded on the returns from increased farm profits—has reduced human disease. Governments have developed antibiotics and vector control technology—justifying funding on the increased efficiency of healthier military forces. Technologies for the control of cholera and yellow fever were developed to facilitate international trade. These approaches continue to provide a potential vehicle for development of innovative financing approaches for the future.

The rate of technological innovation in a society, or in a sector of society, is dependent on the overall economic climate and infrastructure. Specifically, technology embodied in capital equipment and human resources represents an available technology stock for the sector. As in any capital stock, overall rates of investment must be high to allow rapid turnover of the stock. In the current economic environment of the Third World, with many countries just coming out of economic depression, and some of the technological leaders such as Mexico and Venezuela in a debt crisis, it seems doubtful that adequate rates of investment in health system capital can or will be maintained to allow rapid technological innovation.

In general it appears that the low level of investment in health technology in developing countries is in part due to the failure to develop institutions which can appropriate resources and invest. That is, that potential benefits from health technology would justify a higher level of technological investment, but the institutions are not in place that would assure the timely and efficacious use of the technology, nor those that would capture and appropriately allocate the necessary resources.

However, some of the key issues are extremely controversial. Clearly the current U. S. Administration prefers use of private, market oriented institutions where feasible, corrected as necessary. Equally clearly many prefer socialized health systems, using general revenues to support health services as well as heath technology. The social and political implications of these decisions, however, go well beyond the specific impact on health. Moreover, alternative institutions for traditional and paraprofessional health providers appear not even to be considered.

Social Determinants

The relation between culture and development is perplexing. The belief that peasants did not innovate because they were culturally conservative has been discredited as evidence accrues that they do indeed innovate when it is economically beneficial for them to do so. Harrison raises the problem in a particularly evocative way (Harrison 1985), comparing pairs of countries. Why has Costa Rica, with a lesser natural resource endowment, economically surpassed its once richer neighbor, Nicaragua? Why has the Dominican Republic, sharing the island of Hispaniola, surpassed its once much richer neighbor, Haiti? If the answer lies in part in the domains of political stability and effective economic policy, one is left with the underlying question. What is it about the society and its underlying institutions and culture that leads to political stability and economically effective policies?

This question is also at the heart of the study of technology innovation (Elster 1983). Some schools suggest, with Schumpeter, that technology innovation drives economic development. Others believe that innovation is induced by development. All agree that economic development must be attended by technological change. If newly efficient techniques are not substituted for existing ones as the availability and the price of labor and resources change, progress soon will grind to a halt.

In a similar vein, there is a controversy as to whether technological innovators and entrepreneurs produce development or development produces such innovators and entrepreneurs. The latter seems more likely. Anyone who has successfully changed jobs to find a more conducive atmosphere for productivity can testify to the impact of institutional environment on innovation. Similarly, the frequent economic success of migrants from poor countries suggests that it is not their personal abilities or cultural values that determine their relative success—these do not change in transit.

In any society at any time some people, groups, and organizations will be more likely to innovate, and others less likely. It will be possible to predict this propensity, either on the basis of individual characteristics (McClelland 1961), economic characteristics (Cancion 1979), or ethnic or religious origins (Weber 1950). In some situations, innovation is frequently rewarded. If the rewards are sufficiently frequent, innovators will be encouraged to innovate further. Imitators will be encouraged to emulate the innovators. Groups that innovate will prosper, and be likely to attract initiates and imitators. However, if innovators seldom are rewarded and frequently suffer from the innovation, the behavior will tend to be extinguished.

If this analysis is correct, then the central issue is one of organizing the society so that successful innovators are rewarded for their innovations. The technological structure and infrastructure must be sufficiently advanced that the innovator has a reasonable expectation of timely assembly of all the embodiments of his innovation (as well as the resources to innovate—land, labor, capital). Ideally the structure will allow the risk of innovation to be distributed, while the innovator can appropriate a sufficient share of the benefits. Note that an extended family that expects to share equally economic windfalls can be as effective a block to appropriation as an economic policy that favors consumers over producers in the allocation of innovation surpluses. Any society that achieves these conditions, by reinforcing innovating behavior, will produce achievement oriented individuals. These individuals may be Protestants in England in the industrial revolution or Buddhists in today's Japan.

Interpersonal Linkage Patterns

Where do people obtain their goods and supplies, information, resources, respect, legitimation, and values? With whom do people live, work, associate and worship? These are among the prime social facts that condition life. In primary health care we deal specifically with the linkages among practitioner and client, but also with the linkages among practitioners and between practitioners and other members of health care institutions.

Specifically, as one seeks to understand the introduction of computer technology for different groups of individuals—professionals, paraprofessionals, traditional practitioners, families and individuals—one must identify the existing patterns of flow of technology, according to embodiment. Thus, where does the paraprofessional get his equipment, his supplies, his or her education and information. One may utilize the existing linkages for the transfer of computer related technology, in the process changing the nature and legitimacy of the channels. Alternatively one may seek to de-legitimate the existing linkages and establish new ones. In either case, the effect may be far reaching. Thus if one seeks to make traditional curers dependent on the modern health sector for a key source of technology, and de-legitimize the existing religious channels of technological authority, the results may be to undermine other religious linkages. Not surprisingly, such efforts may energize serious opposition on the part of the authorities whose legitimate authority is challenged.

In like manner, the introduction of computers into primary health care, and into the hands of the family itself, requires either changes in the legitimacy and authority of existing linkages for health information, or more fundamentally, a substitution of new for old linkages. The paraprofessional community health worker, provided with a portable expert system, is expected to be a more authoritative source of health advice than older family members, neighbors, traditional practitioners, drug store clerks, and other current source of advice and information. At the least, one hopes that the technology will be sufficiently advanced that the new computerized health professional is a more reliable and valid source of advice than that which he or she is to replace. Nonetheless, one asks how the new linkage is to be legitimated, and how that which it replaces is to be de-legitimated. Similarly, one hopes that the process is not unnecessarily disruptive. Thus one does not wish to diminish the authority of the leader of the family as opposed to that of the outsider, nor to diminish the authority of a member of the same sect as opposed to that of another sect, unless the costs are understood, and the benefits sufficient.

Specific Social Determinants

In the sense of allowing prediction, a number of social variables do serve in our model as determinants of innovations in health-related information technology. It is useful conceptually to disaggregate social determinants into those which occur within developing countries and those which occur in the development of technology for and transfer to developing countries. Similarly, the social factors of importance are different at the source, as compared to those involved in the dissemination and the use of technology.

Urbanization

Urbanization is rapid in all of the developing world. Urban health problems are different from rural, and health services are organized in a different manner in urban than rural areas. Urban populations tend to be better educated, more demanding consumers of health services, better organized to act politically to secure their interests. Communications and infrastructure almost invariably are better in urban than rural areas. Thus it is generally more feasible economically and technically to introduce health-related information technology in urban areas than in rural. On the other hand, experiments in Australia, Alaska and Canada suggest that it may be more beneficial to introduce information technology into health care in rural than in urban areas, since it offers the opportunity to overcome the barriers of distance at low cost.

Modernization

Since the times of Adam Smith and Emil Dirkheim it has been clear that the development involved radically changing the division of labor, allowing workers to specialize, and to increase investment in skills, through the creation of ever more complicated institutions for the articulation of the different functions. In the health field, the modern hospital with specialist physicians represents one extreme of the development continuum, while the self-trained traditional practitioner or self-medicating family member represents the other. Similarly, the research laboratory represents a modern invention which allows specialization in the R&D process, and allows superspecialization of the researchers. In general, more specialized and more articulated systems are more technologically efficient. Specifically, more modern health service systems have a relative advantage in the rate of development and adaptation of microelectronic and microbiological technology. On the other hand, the modern systems have generally had much more attention directed at the improvement of such technological efficiency. As suggested above, attention to formal education and mass media could increase the timeliness of technological innovation in the more traditional sectors.

Clearly, the developing countries in general are moving from traditional practitioner based health-care systems to physician based systems on American and European models. Physician based care dominates the market for upper social and economic classes of the societies, but traditional and paraprofessional based primary health care dominate the lower socio-economic class provision of service. In health delivery systems, as in other sectors, there is a relative dearth of middle class occupations (as compared with developed countries). There are comparatively fewer professional nurses, laboratory technicians, maintenance personnel and other categories, and their functions are carried out by relatively less trained and educated personnel.

To the degree to which computer technology will tend to de-skill the work of physicians, public health and hospital administrators, and other health sector elites, it will challenge the existing division of labor, and tend to decrease their social prestige. To the degree that technology embodied in these systems will tend to require high level knowledge and skills (as in the case of new diagnostic technologies for difficult to treat ailments) it will do the opposite.

Modernization involves some predictable changes in attitudes. Thus, while values vary among societies (Glaser 1970), developed countries are generally oriented toward concepts of equality of access to health services for all citizens as an entitlement. Therefore, they will tend to avoid production of second class technology for poor citizens as immoral. Similarly, developed country populations tend to accept, to a greater extent than LDC's, naturalistic explanations for disease, and technological interventions to prevent or treat diseases. It is common in traditional society to attribute disease to pollution, stemming from some moral transgression (or breaking a taboo). While Western societies take pride in having achieved a more naturalistic explanation of disease, this cognitive conceit has been challenged with regard to health risks even in the United States. (Douglas)

Of particular relevance is the increasing acceptance of the idea of property rights to cultural products which has developed in the West. The ideas that drugs could be patented, or that universities could hold patents, which were controversial fifty years ago, appear much less so today. Similarly, the new technology transfer act in the U. S. allows government scientists to obtain royalties from their government sponsored research, and allows the government labs to keep the license fees. It has become common to copyright software (including the feel and appearance of interfaces) and chip layouts. The resistance of developing countries to these developments may be in part the resistance of the have-nots to the fencing of the commons, but it may also be the result of living in a system that has not developed the social infrastructure to capitalize on the generation of intellectual property.

Professionalization

The professionalization of health occupations is a well established area of medical sociology (Turner 1987:131–156). In Europe the distinctions between physicians, surgeons and apothecaries developed out of class structures, and the guild structure of the middle ages. In the U. S. there remain today conflicts among osteopaths, chiropractors, and physicians. The modern or Western health professions, even when transplanted to the societies of Africa, Asia and the Middle East are accompanied by their institutional framework—associations, licensing, training colleges, and codes of professional conduct. However, they confront categories of traditional practitioners, which have had their own institutional framework.

The various professional disciplines within the health sector carry their own sub-cultural orientations (Pacey 1983:101–103). Thus, physicians are typi- cally acculturated through training and example toward curative medical serv- ices, and frequently are seen by others as relatively unconcerned with preventive or administrative medicine. Sanitary engineers typically are accul- turated to seek engineering solutions to problems, rather than social or politi- cal. In general, such experts have difficulty in subordinating their own expertise in interdisciplinary work, and may withdraw from collaboration into the authority of their own disciplines if challenged. As one seeks to introduce information technology into the health sector, interdisciplinary work becomes inevitable. However, each discipline brings to the process its own particularis- tic values as to the priorities for automation in the sector.

Historically professions have served as organized bodies fostering the interests of their members. A major concern has been the definition of the functions that would be reserved to the members. Obviously the medical profession by increasing the expertise of the physician, and reserving an increasing variety and complexity of functions, has facilitated the improve- ment of the social and economic status of the individual physician. In contrast, the de-skilling of a profession, through division of functions with other groups of workers or through embodying skills in other than the worker, tends to reduce prestige and income of the profession. Thus, it has been suggested, modern pharmaceutical production, packaging, and distribution has de-skilled the practice of the "retail" pharmacist.

Similarly, professions have sought to control entry, and thus to limit supply of services. When service demand falls below supply—as appears to be happening in dentistry in developed countries due in part to preventive meas- ures such as fluoridation—the power of the profession also suffers. Profes- sional groups not only control entry through limiting training opportunities, but also through licensing, and legal enforcement. The profession can act to increase demand, as when the Bolivian union of unemployed physicians demonstrated for increased public employment, or when professional associa- tions lobby for public subsidies for fees-for-service or third party insurance.

The professional associations also seek to influence the institutionaliza- tion of health services. Medical associations, for example, are generally very active in decisions on the socialization of medicine, and on the structure of health financing systems and organizations. In like manner, medical faculties are active in the planning and management of medical schools, and medical researchers have attained considerable power in the institutionalization of medical research.

Innovations in information technology potentially would allow a radical restructuring of the health professions. For example, computerization of phar- maceutical information would offer the possibility to professionalize pharma- cists at the expense of physicians, allowing the pharmacist to take a more active role in prescribing and monitoring drug usage. Similarly, primary-care expert systems and computerized diagnostic instruments would offer the possibility of professionalizing paraprofessional occupations at the expense of physicians, who might then play more of a supervisory role or focus more on secondary and tertiary care. In some areas, such as the clinical laboratory, computer technology may be involved in a complete restructuring of work and realignment of professional function. Moreover, new professions may evolve to carry out new functions inherent in the new information technology, as occurred with radiology in the past.

The social fact of the existing pattern and power structure of professionalization will surely condition the pattern of introduction of the technology, and of professionalization and de-skilling of occupations. Thus it is hard to see how religiously based traditional practitioners in the Middle East or tribal societies will overcome resistance from physicians in their societies to appropriate information technology to their interests. Similarly, it is difficult to see how midwives will get equal access to information technology as compared with obstetricians.

A special characteristic of information technology in health is the close relation of the technology to information science and medical research. Science, as a result of European traditions, is generally regarded as a gentlemanly activity directed at the attainment of knowledge, and characterized by peer review and open exchange of information. In contrast, technology is a trade related effort to develop productive efficiency, and characterized by ownership of information. Science is the domain of the disciplinary department in the university, while technology is the realm of the sectoral department in government or industry. As microelectronics and knowledge engineering move from science to technology, action is moving from academe to the private sector, information from professional journals to patent applications, and attitudes from cooperation to competition.

Viewing researchers as social individuals, their training involves not only the assimilation of knowledge of the subject and the skills of the scientist, but also the acculturation into the values of the group. They tend to work on problems which their colleagues agree are appropriate and important. Since the vast majority of researchers are concerned with the problems of developed countries, it is not surprising that young Third World scientists are often acculturated to share this orientation. Similarly, American scientists interested in Third World health technology may have serious concerns that this will hinder their careers and pull them out of the "main stream".

In developing countries, the scientist is faced with a peer group that does not do research, and faces pressures to conform to that reality (Choudhuri 1985). In countries that value teaching more highly than research, that confer more prestige upon the humanities, law and management than upon medicine and engineering, that simply do not have the institutional structure to support highly productive research, this professional orientation may be natural. It results, however, in relatively low rates of technology innovation.

Dualism

A useful abstraction for the discussion of developing country social structure is "dualism". Thus in many countries there are significant differences and social distances between a modernizing, relatively affluent, urban population and more traditional, poor, and rural populations. The differences are emphasized in those countries in which tribal cultures dominate rural areas, or those in which the educated elite use a foreign language and look toward a metropolitan center. In general, modernization of health technology has been faster in the urban elite sections of the population. One also may generalize that where traditional and modern health practices clash, there is significant potential to damage traditional value structures. In part this comes simply from the reduction of the legitimate authority of traditional sources of health technology, and in part from modern deprecation of traditional practices such as circumcision and acupuncture that have social as well as physiological impacts.

Cultural dissonance can be expected between developed country sources and developing country users of technology. Technology tends to carry with it to the new society attributes from the society in which it was developed. For example the prevailing tendency for technical assistance experts and donor agencies to seek to model developing country institutions on the pattern of their home country has been termed "homomorphic institutionalization" (M. Ivory pers. com.). The result of the cultural distance can be delays and difficulties in technology transfer; the result of the pressure to make the developing country conform to foreign norms can be cultural dissonance or disruption.

Gender Biases

Preference for males is a common value feature of many societies. It is so pronounced in some that there are very high infant mortality rates among girls, and adult sex ratios biased toward males. The situation is complicated in some countries by social barriers between males and females so that, for example, male practitioners cannot attend female patients. The social distance between males and females, and the social inferiority of females in some countries interferes with the transfer of technology to female users, with the diminishing potential role of the female as a health service provider, and as a researcher.

There is also a common pattern of differentiation in the technological roles of men and women. Thus Pacey writes:

> It is notable that the traditional division of labor between man and women casts men as the makers of tools and equipment, thereby giving them a

greater interest in the 'expert sphere' of technology, while women are often most directly concerned with the end-use of equipment or energy, and with meeting basic needs. Thus women tend to experience technology less as making things and more in terms of 'management of process' leading to a very distinctive outlook. The importance of this is not usually recognized because of the habit of regarding women's traditional roles as service activity, subsidiary to the more serious business of wealth creation (Pacey 1983:101-102).

Socio-Economic Class

The nature of socio-economic class distinction differs from culture to culture, but the fact of class difference is universal. Moreover, the relative strengths of the different classes, and the degree of class difference and class conflict appear to be important determinants of the pattern and rate of technological innovation in the health sector.

While formal educational levels are generally lower in developing countries in developed countries, higher classes generally have more education than lower classes in their own societies. In addition to providing information and skills which are important to technological innovation, formal education generally imbues the student with positive attitudes toward science and scientific epistemology. It also tends to legitimate modern institutions as authorized sources of technological information—such as doctors, hospitals, books, and mass media. One may expect that positive values toward modern education will be associated with other positive values toward modern health technology.

Generally urbanization, modernization, and professionalization are all correlated with class. Thus higher socio-economic class is related to urban residence, modern outlook, and membership in professions with higher socio-economic status. Moreover, patients with higher socio-economic status are likely to seek primary health care from relatively high socio-economic status providers. Thus the conflicts over skilling and de-skilling of professions, or over resources for one or anther type of health services are likely to engender and be engendered by other conflicts among different socio-economic classes.

Political Determinants

Partisans of the extreme left suggest that technological innovation in health in developing countries is an effort of the extreme right to ameliorate conditions that promote social unrest without dealing with the underlying causes if socio-economic inequity. Partisans of the extreme right suggest that the extreme left may resist microelectronics technology as putting control of information into the hands of the many, in ways incompatible with the needs of centrally planned societies. It appears best to avoid such partisan arguments, which in any case are difficult to support with evidence. Rather, it is urged that interested parties in health technology do organize, and that so organized they exert political influence which is an important determinant of the realization of inherent technological potentials.

The political needs differ for different technological systems. The private sector, for example, needs merely a permissive environment to develop and transfer technology since it has ways to appropriate the resources it needs. On the other hand, public institutions need positive support and subsidy on a continuing basis. The lack of continuity in support is a particular hazard in health technology development and testing, given the long time delays and high costs. It is difficult to develop the personnel, instrumentation and organizational capacity to do state-of-the-art health technology development in the fields of microcomputer and microelectronics technology.

Ideologies

The political debates involved in the development and dissemination of health technology are deeply influenced by ideological positions. For example, the development assistance program of the U. S. has been primarily a result of the liberal political ideology of the U.S., which in turn comes from some deeply held beliefs: "1. change and development are easy, 2. all good things go together, 3. radicalism and revolution are bad, and 4. distributing power is more important that accumulating power." (Peckenham 1973). The dissonance that this position creates when confronted by the ideological position of the centrally planned economies or the Third World is significant.

The U. S. tends to have, like some political groups in the Third World, a technocratic ideology. It is perceived that policies should be made on technical grounds, that technocrats should manage government agencies, and that technological solutions are frequently possible for social and economic problems. In international *fora*, such as the World Health Assembly, U. S. spokespersons frequently seek to divert discussions into "productive" technical areas, and away from "unproductive" political discussions.

Nationalism, the free-market, and central-planning have been central ideologies in the North-South debates on technology transfer. To some degree these debates appear to have moderated in the 1980's as compared with the 1970's. However, as the U.S. seeks to reestablish an acceptable trade balance, and as the Third World moves from depression and debt crisis to renewed concerns for development and export promotion, frank economic confrontations seem more likely. (While there is also concern about the transfer of high technology from the West to the East, this does not seem to be a major source of ideological concern in the area of health technology.

Power Relationships Relative to Sources of Technology

As medical research has become a big business, it also has become a political issue. Industries (pharmaceuticals, medical devices), universities, bureaucracies (government regulatory and research agencies), and public interest groups exercise political influence actively and deliberately. In many ways the results are predictably similar to those in other sectors.

The pattern of the U. S. National Institutes of Health has been important in development because of its own efforts and because of the role the NIH has played as a model. The NIH is unusual in that it has deliberately sought to build a political constituency in the medical and university community, and has accepted a research grants approach to that end. In the search for professional support they have given unprecedented power to the researcher. "NIH is the only major dispenser of Federal funds in which the responsible full-time officers cannot legally make a grant without the approval of part-time committees representing the beneficiaries." (Price 1973)

The political support for public support of medical research in the developed countries also has benefited from the activities of public campaigns, prototypically that of the March of Dimes. (Strickland 1972) While mass media campaigns do collect a significant amount of financing, they also educate the public and raise political support for the funding of research. It is not coincidental that the U. S. NIH is organized into disease specific institutes, each of which can count on its public support group, rather than on disciplinary lines. Similarly, the public support in the U. S. for "child survival" and for "food for the world" has been instrumental in obtaining resources not available for "international health".

The allocation of support for health technology development among countries is also subject to international political activity, as illustrated by the debates over support for the UNIDO biotechnology center, or over the UNCSTD proposal of an international fund for science and technology for development.

Power Relationships Relative to Users of Technology

Not only the outcome of the general political actions in developing countries, but the health specific political controversies will affect the kinds of technology used and will to some extent determine the beneficiaries. For example, the relative success of urban economic elites, manufacturing workers, the military, and similar groups in obtaining priority access to medical resources in most countries is well documented.

Clearly the division of labor in the medical sector is in part the result of the exercise of political power by the occupational groups involved. In general, the licensing of physicians and the legal requirement that physicians carry out specific responsibilities is a major tool in maintaining the unequalled professional authority of the physician. Similarly, political power can and is exercised within hierarchical organizations such as ministries of health and hospitals.

Sexual politics have become relatively visible in the developed world, and may become more so in developing countries. Thus the ideal of empowering women may be evidenced in "comparable worth" campaigns to secure pay for occupations predominantly staffed by women that are deemed commensurate in responsibilities and training to other male dominated occupations. In the case of patient care, there are campaigns to give female patients more equal status in the patient-physician relationship (Arney 1982).

Power Relationships Relative to Technology Transfer

In many countries the pharmaceutical industry appears to have succeeded in co-opting the medical profession, medical schools, and even regulatory agencies. Such co-optation comes from the extensive system of personal contacts maintained by the companies with these groups, through use of corporate resources to support professional activities, and through other routes.

The political aspects of international technology transfer include those involved in the support and regulation of health related MNC's, the political aspects of control of UN agencies (WHO, PAHO, UNICEF especially), and debates on human resource issues such as the brain drain and its relationship to foreign training for Third World health professionals.

The international political aspects appear to be especially visceral in areas in which ethical and moral values are at play, such as contraception and abortion (Djerassi 1981). An alternative issue of considerable import has been the infant formula controversy, in which only the U.S. has opposed World Health Assembly votes for a code of conduct—based on a free enterprise philosophy which down-plays the continued concerns for perceived dysfunctions of formula marketing in developing countries.

Colonialism and Neo-Colonialism

European and United States' influences have been pervasive over a period of centuries on the health systems of the developing world. They leave a legacy which is obvious, and likely to be profound in the transfer of health-related information technology.

In Latin America, Spanish conquerors brought their own system, in which:

> there existed three types of medical assistance: for the powerful, for the artisans and bourgeois, and for the poor. The proto-physicians, graduated from the most important universities, attended kings, princes and nobles. Their fees were elevated, their honors ostentatious, and their obligations as important as delicate. For well-trained physicians and surgeons, the bourgeois clientele was increasing. Consultation was in the home, the professional very proper in his behavior and attention, and the fees very elevated. An adequate remuneration permitted the birth of the 'physician in charge' and liberal medicine. Finally, the care of the poor remained in the hands of traditional healers and barbers. In the cities the poor could seek refuge and die in hospitals and hospices put at their disposition by municipal charity (Garcia 1972:392,393).

The historic roots of Latin American medical systems are clearly to be found in these Spanish antecedents. Moreover, the Galenic concepts of the causality of disease brought on by imbalances of the four humors of the body—hot, cold, moist, dry—are still reflected widely in Latino beliefs about "hot" and "cold" foods and medical conditions.

More controversial is the suggestion that the colonial experience transformed and deepened the medical superstitions in traditional societies:

> With Christianity, so it seems to me, the missionaries also introduced magic, or *magia* as it is called in the Putamayo today in reference to power that stems from a pact with the devil. The missionaries believed firmly in the efficacy of sorcery, which they supposed Indians to be especially prone to practice on account of their having been seduced by the devil (Taussig 1987:142,143).

In North Africa, European medicine replaced Arab as the dominant paradigm in the 19th century. Thus a cholera treatise written in Turkey by Mustafa Beheet in 1831 was widely distributed in Tunis. Beheet's cholera treatise represented a technological response to a new epidemic through the Ottoman paradigm. Similarly, in the 1850's as Tunisia's Ahmed Bey opened and expanded hospitals and planned medical schools in Tunis, the chief doctor of Egypt, Muhammed Ali, Clot Bey, sent Ahmed a shipment of medical books in Arabic and a letter describing Egyptian achievements in health care (Gallagher 1983). By the end of the century, however, "Europeans dominated the medical establishment. Arabic medical practitioners were officially discredited. The transition to European medicine, like economic reversal, was a result more of the process of colonization than of the struggle with epidemic disease."(Gallagher 1983, 101) The European technology was not necessarily more effective than the Arab.(Gallagher 1983, 3-13) At the time both cultures abounded in animistic theories; both cultures in the 18th century were innovating, especially in responses to new challenges. In fact in Tunis, Arab authorities introducing quarantine to control epidemic diseases were opposed by Europeans (Gallagher 1983, 58,59).

In Asia, the development of the Peking Union Medical College with the assistance of the Rockefeller Foundation stands out, transforming a previous missionary school into a center of excellence in the training of physicians. The curriculum was based on the Johns Hopkins model of science based medical practice. The foundation rejected the idea advanced by Abraham Flexner of training for a simplified medicine, more practical in terms of the needs and resources of the Chinese. In fact they sought to train a cadre of scientifically informed physician-leaders, who would in turn transform medicine in China towards the pattern of United States and European practice. This practice Rockefeller was to replicate on many continents. While one can little doubt that the principals in the endeavor were imbued with humanitarian motives, the PUMC project was explicitly defended on the basis of using medical technology as a wedge to open China to U.S. commercial interests (Brown 1982). The colonial legacy includes medical systems strongly influenced by Western models, a historical dependency of developing country institutions on innovations from the North, and an increasing distrust of the social consequences of too facile a reliance on such Western innovations.

Positive and Negative Impacts

One recalls with some humility the expert who at the turn of the 20th century correctly predicted the magnitude of the investment necessary to build the telephone system for the U. S., and then stated that such an investment was clearly impossible, and predicted the telephone system would never be built. The printing press was promoted originally by those who believed it would promote Christian morality by increasing the reading of scripture. Despite the obvious dangers, the following paragraphs project some likely eventualities.

Computers will be widely introduced in developing country primary health care systems in the next decades. The radical reduction in cost of hardware and general purpose software will make information technology significantly more affordable. Applications will include expert systems for patient diagnosis, triage and treatment, as well as a variety of administrative and communications functions.

However, the introduction of information technology embodied in hardware will be more complete than that embodied in software. There will be a considerable lag in developing a cadre of trained personnel to utilize the technology, and of reforming organizational structures and processes to make maximum use of the potential in informatics. Applications closely related to those of developed countries will lead, such as financial management, laboratory automation, or patient record management, and applications unique to developing countries will lag, such as applications to tropical disease or to paraprofessional health workers. Lacking ways to subsidize the development of theory and data, there will be under-investment in unembodied technology specific to the introduction of health-related information technology in developing countries.

The potential benefits from the application of microcomputers to tropical disease, and from improving health services for the poor are not likely to be fully achieved. Because researchers are concentrated in developed countries and because the scientific community is focused on problems of the developed countries, the potential for development of microcomputer and microelectronics technology for poor people in developing countries will probably lag that of technology for more affluent people and countries. Similarly, due to the lesser ability to pay and the less efficient institutions to appropriate benefits from such technology, the world will probably under-invest in tropical disease and similar technology. The increased availability of high cost technology, affordable only by elites, will exacerbate the internal competition for health care resources, and the power advantage of the elites will result in increased inequity in distribution of services and resources.

Physicians and hospitals probably will be more successful in appropriating microcomputer and microelectronics technology than paraprofessionals and primary health services, and therefore will exacerbate a trend toward high cost hospital-based medical services. The potential in computer technology to de-skill health care probably will be resisted effectively by physicians, due to their political and organizational authority. The reduced market for information technology for paraprofessionals and primary facilities, and the cultural bias of the R&D community, will probably militate against such applications being quickly or fully developed. As a result of these tendencies the technological potential available now to reduce morbidity and mortality at low cost will not be fully realized.

The potential to increase self-reliance of individuals and families, or to increase the role of traditional practitioners probably will be less fully realized than the potential to increase the role of the physician/hospital based sector. This is due to the fact that the general public will have less incentive, and is served by a less efficient technological system than the health professions, and due to the fact that traditional technology systems are less efficient innovators than the modern medical technology system, and that traditional practitioners are provided only partial, controlled access to the modern system. The results may include greater authority for the medical profession, and greater dependency to patients, as well as more distance in values between patient and practitioner than would otherwise occur. In like manner, in most developing countries men probably will have greater access to health-related information technology innovations than women, reflecting and reinforcing occupational barriers in the health industry to women and the dependency of women in the society.

The class distinctions which currently characterize the provision of primary health care will be accentuated. As physicians and the more affluent health service institutions more effectively appropriate computer technology to increase the efficacy and range of their services, the benefits of modern, physician and hospital-based services will increase. Elites, either through private (prepaid practices or fee-for-service) or more likely through control of specialized quasi-public (military and social security) institutions will then compete for additional primary care resources. Given their economic and political power, and their alliance with the more powerful professional and health industry lobbies, these elites will be relatively more successful in the competition for resources than poorer, and more marginalized groups.

The potential to improve decision making and to make health services more responsive to health needs that is inherent in computer technology for primary health care probably will not be achieved in developing countries. This is a function of both the institutional rigidities inherent in political, bureaucratic and legalistic systems, and attitudinal systems that promote other values over nominal rationality.

Recommendations

If the negative outcomes suggested above are to be reduced, the capacity of the developing countries to develop, adapt, introduce and analyze the impact of computer technologies in primary health care should be improved. This would increase the rapidity of development of appropriate health-related information technology for the special needs of developing countries. Social scientists, who have the scientific training to analyze the economic, social and political determinants of technology transfer should be intimately involved in technology assessment, and in the specification of new institutions to improve the generation, dissemination and use of health related information technology. A wide variety of people from the societies involved, who embody the relevant moral and ethical values should be involved in health technology policy debates in their own countries; specialists in the relevant philosophical and religious traditions also should be involved with appropriate authority.

More specifically, the institutional basis for the development and dissemination of health-related information technology in developing countries should be improved. Emphasis should be given to institutional forms with which to subsidize research in computer technology for preventive measures and for paraprofessional based primary care. Such institutional efforts should include: (a) institutions directing professional attention toward these problems such as new journals and series of scientific conferences; (b) new financial institutions, such as new consortia of donors for technology development, or financing treaties among developing countries for joint support of international centers or for treaty-defined internal research, accompanied by specific funding for technology exchanges.

There is also a need for donor subsidies for projects to promote the introduction of computers in primary health care systems which emphasize paraprofessionals, and for subsidies to encourage the introduction of the technology in traditional practice and in non-medically based practice such as that of popular pharmacies. Such subsidies can be in the form of: (a) demonstration projects, funded through Private Voluntary Organizations, government, or the private sector; (b) subsidies for development of software, courseware, and applications of microcomputers for these intermediaries; (c) development of training curricula and institutions to use computer systems in the training of low and mid-level manpower, especially in training of midwives, Ayurvedic practitioners, and pharmacy clerks. Finally, there is a need to subsidize the development and application of computer technology for the use of families and individuals in developing countries in meeting their own primary care needs.

These are daunting recommendations. The strengthening of science and technology capacity in developing countries, for example, has been a major focus of development efforts for decades. Yet in large regions and for long periods, the result has been decay in scientific and technological capacity rather than growth. Nonetheless, the potential power for both benefit and damage inherent in the current technological revolution creates a moral imperative to redouble efforts to develop appropriate health technology for the majority of the people of the world. Moreover, the dignity of the civilizations of the Third World, and the need for cultural alternatives to those of the First World demand that the people of the Third World exercise moral leadership to control their own technological destinies.

Conclusions

Computers have not yet been widely used in primary health care in developing countries. However, it seems likely that in the course of the next decades, there will be a major dissemination of the technology to carry out a wide variety of functions. Already we have moved in this direction. History suggests that in both health systems and information technology, it becomes increasingly difficult to undo that which has been done; the technological, institutional and social processes that are unleashed have their own momentum. As time passes, while we perceive through experience more fully the implications of our past actions, return to previously viable options becomes more difficult.

In the case of medicine, the need to develop social institutions to finance the escalating costs of hospital care needed to apply the rapidly developing technology of secondary and tertiary medical care in the middle third of this century led to supportive social decisions: tax subsidies to fee-for-service, third party insurance, prepaid health plans, socialized medicine (Zeltan 1979). These new institutional arrangements had dramatic effects on the resulting hospital based care system. Moreover, once the systems have been fully institutionalized, their beneficiaries constitute enormous and powerful lobbies for maintenance of the forms chosen. The very existence of the institution provides a vehicle to coordinate the support, while the lack of the alternative not chosen limits the potential for its proponents to organize effectively.

It is not difficult to predict that, as the functions allocated to the health professions are restructured, based on automation of information functions, strong social pressures will be engendered to maintain the resultant professionalization structure. If, for example, a developing country seeks to strengthen greatly the role of medical auxiliaries through the development and use of hand-held, expert systems and low-cost, "smart" instruments, then the resulting large number of such practitioners will become a strong lobby for their profession, as may their patients. On the other hand, if technology is used primarily to greatly expand the coverage provided by primary care physicians, through use of improved communication and automation of information activities, the physicians will become a still more powerful lobby (as they did with the widespread use of telephone and automotive technology in the U. S. which greatly increased the efficiency of the individual practitioner).

Insofar as we can foresee the future implications of actions, it might seem that we should seek those utilitarian courses of action that are expected most nearly to achieve our goals. However, we live in a world of ethical relativism. Certainly the values that control ethical choice differ from developing country to developing country, and from donor to recipient country. The problem is real. Consider a U.S. consultant, funded by a donor agency, negotiating with a host country official about the introduction of computers into a maternal child care system. Such a person may face dramatic value conflicts about the value of human life, the role of women, the importance of intra- and inter-professional courtesies, and the intrinsic value of rationality. Moreover, the consultant will not only face the differences of his or her personal values with those of the other negotiator, but also with values represented by the donor agency, the host agency, and the client population.

Moreover, as Robert Textor points out we are tempocentric. (Textor 1984) Thus medical ethics, as codified by the medical profession, have changed radically over time (c.f. Chapman 1984). Only relatively recently (1957) in the United States, for example, has the primary ethical responsibility of "placing the patient's interests before all else in the medical relationship" begun to be recognized in the code of ethics of the American Medical Association(122). (Note that this utilitarian objective is still not the paramount imperative in Soviet, deontological, medical ethics, which stress ideological, scientific and intellectual preparation of the practitioner(130,131). This of course may be a reflection of the improvement of technology which only recently has offered the physician a realistic probability of contributing to the recovery of the average patient encountered.

It is in this context of a new technological potential that we focus on primary health care. Can we not seek, in this one limited domain, to put first the authority of the competent individual over his or her own health, and next the right of the individual to the most complete physical, mental and social well-being attainable. Let us then seek, in this limited domain, to build those institutions and engender those processes that will encourage the technological potential inherent in the microchip to be fully utilized to solve the visible, painful, and pressing problems of death, disease, disability and human suffering. Let us recognize that in the right to health, the only equity is in equality, and thus emphasize the poor who suffer most.

References

Arney, W.R. 1982. *Power and Profession of Obstetrics*. Chicago: The University of Chicago Press.

Beniger, J.R. 1986. *The Control Revolution: Technological and Economic Origins of the Information Society*. Cambridge, MA: Harvard University Press.

Beverly, J. 1984. Technology transfer in water supply and sanitation. In Summary of AID office of Health Technology Workshop. WASH Working Paper No. 25, prepared under AID Contract No. AID/DSPE-C-0080. Washington, D.C..

Botero, C. 1972. Drugs and dependency in Brazil: An empirical study of. Ph.D. Diss. Ithaca, NY: Cornell University.

Bowers, J.Z. 1977. *An Introduction to American Medicine-1975*. U.S. Dept. of Health Education and Welfare Publication No. (NIH) 77-1283. Washington, D.C.

Broadman, J.Z. January, 1988. Using management tools to get the most from information technology in development. Presented at the National Academy of Sciences Workshop on Microcomputer Policy for Developing Countries. Washington, D.C.

Brown, E. 1982. Rockefeller medicine in China: Professionalism and imperialism. ed. Robert F. Arrove. In *Philanthropy and Cultural Imperialism: The Foundations at Home and Abroad*. Bloomington: Indiana University Press.

Cancion, F. 1979. *The Innovator's Situation: Upper-Middle-Class Conservatism in Agricultural Communities*. Stanford, CA: Stanford University Press.

Cartwright, F.F. 1967. *The Discovery of Modern Surgery*. London: Arthur Barker Ltd.

Chapman, C.B. 1984. *Physicians, Law and Ethics*. New York: New York University Press.

Choudhuri, A.Rai. August 1985. Practicing western science outside the west: Personal observations on the Indian scene. *Social Studies of Science*. 15(3): 475-505.

Cilingroglu, A. 1975. *Transfer of Technology for Pharmaceutical Chemicals*. Paris: OECD. Organization for Economic Cooperation and Development.

Coleman, J.S., E. Katz, and H. Manzel. 1957. The diffusion of innovations among physicians. *Sociometry*. 20: 253-270.

____. 1966. *Medical Innovation: A Diffusion Study*. Indianapolis: Bobbs-Merrill Co.

Comroe, J.H. 1976. Lags between initial discovery and clinical application to cardiovascular pulmonary medicine and surgery. Appendix B in *Report of the President's Biomedical Research Panel*. Washington, D.C.

Crain, R.L., E. Katz, and D.B. Rosenthal. 1969. *The Politics of Community Conflict: The Floridation Decision.* Indianapolis: Bobbs-Merrill Co.

Daly, J. April, 1986. "Pan American Health Organization develops computer translator," *Poplac News.*

Developing World Industry and Technology, Inc. 1979. Changes in the terms and conditions of technology transfer by the pharmaceutical industry to newly industrializing nations over the past decade. Final report on AID contract no. AID/DSAN-147-697.

Djerassi, C. 1981. *The Politics of Contraception: Birth Control in the Year 2001.* San Francisco: W.H. Freeman & Co.

Douglas, M., and A. Wildavsky. 1982. *Risk and Culture.* Berkeley: University of California Press.

Eden, M. 1986. Currents in Computer Technology and their Effects on Biomedical Instrumentation. Paper presented at the NAS-JNICT Seminar on Microcomputers in Development, Lisbon.

Educational Commission for Foreign Medical Graduates. 1984. *Education of the physician: International dimensions.* Washington, D.C.

Elster, J. 1983. *Explaining Technical Change.* Cambridge: Cambridge University Press.

Fraser, R.W. 1979. *Guidelines for analysis of socio-cultural factors in health.* U.S. DHEW Publication No. (PHS) 79-50083. Washington, D.C..

Frerichs, R.R. June, 1987. *Young Child Cluster Survey: A Report on the First Rapid Community-Based Survey in Burma.* San Francisco: Western Consortium for the Public Health, Inc. mimeographed.

Gallagher, N.E. 1983. *Medicine and Power in Tunisia, 1780–1900.* Cambridge: Cambridge University Press.

Garcia, J. 1972. Translated from *La Educacion Medica en la America Latina.* Washington, D.C.: Pan American Health Organization

Glaser, W.A. 1970. *Social Settings and Medical Organization: A Cross National Study of the Hospital.* New York: Atherton Press. Inc.

Harrison, L.E. 1985. *Underdevelopment is a State of Mind: The Latin American Case.* Lanham, MD: Madison Books.

Hashino, S. April 1986. Recent Progress in Robotics. In ASEAN Committee on Science and Technology, *Proceedings of the First ASEAN Science and Technology Week Conference.* 2:910–923.

Heise, D.R. 1981. The microcomputer revolution? Technical possibilities and social choices. *Microcomputers in Social Research.* Beverly Hills, CA: Sage Publications. (Special issue of *Sociological Methods Research* 9(4): 395–437.)

Hirsch, P. Sept. 1975. Organizational effectiveness and the institutional environment. *Administrative Science Quarterly.* 20: 327–344.

Hopkins, D.R. 1983. *Princes and Peasants: Smallpox in History.* Chicago: The University of Chicago Press.

Jadlow, J.M. 1976. An empirical study if the relationship between market structure and innovation in therapeutic drug markets. Final Report, NSF Grant No. RDA75-21075, Stillwater, OK: Oklahoma State University. mimeographed.

Kastner, J.K., C.R. Dawson, S.M. Weiss, K.B. Kern, and C.A. Kulikowski. 1984. An expert consultation system for frontline health workers in primary eye care. *Journal of Medical Systems.* 8(5): 389–97.

Kraemer, K.L., and J.L. Perry. May/June, 1979. The federal push to bring computer applications to local government. *Public Administration Review.*

Kraemer, K.L., and J.L. King. 1984. National Policies for Local Government Computing: an Assessment of Experience in Ten OECD Countries. *IRAS* 2: 133–147.

Lukas, T.A. 1978. *Study proposal: The determinants of innovation in medical devices.* Washington, D.C.: U.S. Congress. Office of Technology Assessment. mimeographed.

Magee, S.P. 1977. Information and the multinational corporation: An appropriability theory of direct foreign investment In *A New International Economic Order: The North South Debate.* ed. J.N. Bhagwati. Cambridge, MA: MIT Press.

Mansfield, E. et al. 1971. Innovation and discovery in the pharmaceutical industry. In *Research and Innovation in the Modern Corporation.* New York: Norton and Co.

McClelland, D. 1961. *The Achieving Society.* Princeton: Van Nostrand Co.

McCraine, N.E. 1976. "The multinational pharmaceutical industry: Implications for national development." M.A. Thesis. University of Texas at Austin.

Munasinge, M. January 1988. The Role of Computers and Informatics in Developing Countries: Issues and Policy. Paper presented at the National Academy of Sciences Conference on Microcomputer Policy for Developing Countries. Washington, D.C.

National Academy of Sciences. 1973. *U.S. International Firms and R,D & E in Developing Countries.* Washington, D.C.

National Academy of Sciences. 1986. *Microcomputers and Their Applications in Developing Countries.* Boulder, CO: Westview Press.

National Academy of Sciences. 1979. *Medical Technology and the Health Care System: A Study of the Diffusion of Equipment-Embodied Technology.* Washington, D.C.

National Academy of Sciences. In Press. *Policy Issues in Microcomputer Applications for Developing Countries.* Boulder, CO: Westview Press.

Organization of American States. 1982. *Biopharmaceutical evaluations: Final Report.* Washington, D.C.

Pacey, A. 1983. *The Culture of Technology.* Cambridge, MA: The MIT Press.

Peckenham. 1973. *Liberal America and the Third World: Political Development Ideas on Foreign Aid and Social Science.*

Perrow, C. 1986. *Complex Organizations: A Critical Essay.* (Third Ed.) New York: Random House.

Piachaud, D. 1979. The diffusion of medical techniques to less developed countries. *International Journal of Health Services.* 9(4): 629–643.

Pillsbury, B.L.K. 1978. Traditional Health Care in the Near East. U.S. Agency for International Development, Contract No. AID/NE-C-1395. Washington, D.C.

Pool, I. de Sola. 1984. Tracking the flow of information. *Science.* 221(4611): 609–613.

Population Reports. May, 1980. Traditional midwives and family planning. 1(22).

_____. 1985. Contraceptive social marketing: lessons from experience. 1(30).

Price, D. de Solla. 1986. *Little Science, Big Science..and Beyond.* New York: Colombia University Press.

Price, D.K. 1973. *A political hypochondriac looks at the future of medicine.* Washington: National Academy of Sciences.

Rogers, E.M. 1983. *Diffusion of Innovations.* Third Ed. New York: The Free Press.

Rothman, J. 1980. *Using Research in Organizations: A Guide to Successful Application.* Beverly Hills: Sage Publications.

Scholarly Communication. 1979. *Report of the National Enquiry.* Baltimore: Johns Hopkins University Press.

Schware, R. 1987. Software industry development in the third world: Policy guidelines, institutional options, and constraints. *World Development.* 15(10/11).

Shyrock, R.H. 1980. *American Medical Research Past and Present.* New York: Arno Press. (Originally published in 1947).

Starr, P. 1982. *The Social Transformation of American Medicine.* New York: Basic Books.

Steele, W.W., and S.A. Oesterling. 1984. *The graduate education of foreign physicians in public health and preventive medicine: The role of United States teaching institutions.* Educational Commission for Foreign Medical Graduates.

Strickland, S.P. 1972. *Politics, Science, and Dread Disease.* Cambridge, MA: Harvard University Press.

Taussig, M. 1987. *Shamanism, Colonialism and the Wild Man: A Study in Terror and Healing.* Chicago: The University of Chicago Press.

Textor, R. 1985. Shaping the Microcomputer Revolution to Serve True Development. In *Microcomputers for Development: Issues and Policy.* ed. Munasinghe Mohan, D.M. Dow, and Jack Fritz. Columbo Sri Lanka: National Computer Council.

Textor, R.B. et al. 1983. *Austria 2005: Projected Sociocultural Effects of the Microelectronic Revolution. 45–40.* Orac Pietsch: Distributed in U.S. by Stanford University.

Tornatzky, L.G., J.D. Eveland, M. Boylan, W. Hertaner, E. Johnson, D. Roitman, and J. Schneider. 1983. *The Process of Technological Innovation: Reviewing the Literature.* Washington, D.C.: National Science Foundation.

Turner, B.S. 1987. Social organization of medical power: Professions, knowledge and power. In *Medical Power and Social Knowledge.* Beverly Hills, CA: Sage Publications.

U.S. Congress. Office of Technology Assessment. 1976. Development and diffusion of medical technology. Appendix A in *Development of Medical Technology: Opportunities for Assessment,* Washington, D.C.

_____. 1986. *Technology transfer to the Mid-East.* Washington, D.C.

U.S. Department of Health, Education and Welfare. 1971. Planning for creative change in mental health services: A distillation of principles on research utilization. DHEW publication No. (HSM) 71-9060. Kent, OH: Kent State University.

U.S. Department of Health, Education and Welfare. 1976. Appendix A: The place of biomedical science in medicine and the state of the science. *Report of the President's Biomedical Research Panel.* U.S. DHEW Publication No. (OS) 76–501. Washington, D.C.

Weber, M. 1950. *The Protestant Ethic and the Spirit of Capitalism.* New York: Charles Scribner & Sons.

Zaltman, G., R. Duncan, and J. Holbek. 1973. *Innovations and Organizations.* New York: John Wiley & Sons, Inc.

Zeltan, R.A. May 1979. Consequences of increased third party payments for health services. *The Annals of the American Academy of Political and Social Sciences.* 443.

12

A Decision Support System for Village Health Workers in Developing Countries

*G. Porenta, B. Pfahringer,
M. Hoberstorfer, and R. Trappl*

Introduction

"Health for all by the year 2000" is one of the main goals the World Health Organization (WHO) set forth in its program. This effort would include helping underdeveloped countries to establish a sufficient standard of public health care. Currently, though, the gap between health care in industrialized countries and that in underdeveloped countries seems to be increasing instead of decreasing. The majority of the ailments that people in developing countries suffer from could in principle be effectively treated by simple strategies. However, both knowledge about treatment plans and resources are commonly scarce in countries where they are needed most urgently.

Therefore, the salient problem in providing adequate health care is the distribution of knowledge and not a knowledge deficit. For centuries, tutoring on a personal basis and transmission of knowledge in books have been the

This work was supported by the Austrian Federal Ministry for Science and Research and the Austrian Labour Union. It has been discussed by the members of the project "AI-Based Systems and the Future of Language, Knowledge, and Responsibility in Professions" within the COST 13 framework of the European community. We gratefully acknowledge the help of Dr. Lindmaier in the design of the knowledge-base covering the problem area "skin." Ms. Glatzer provided valuable comments from a nurse's perspective. Dr. Lang helped with her medical expertise during the final evaluation and debugging of the system. Reprinted with permission from Applied Artificial Intelligence 2:47–63, 1988 Copyright © 1988 by Hemisphere Publishing Corporation.

pillars of knowledge transfer. Personal tutoring correctly performed provides active transfer of knowledge, whereas books present passive blocks of information waiting for the student to act according to rules without any feedback on performance.

With the advent of knowledge-based systems, the computer seems to emerge as another viable way of transferring knowledge in a more active form by prompting medical personnel for appropriate responses. Especially for developing countries, knowledge-based decision support systems on rugged and portable personal computers independent of AC power supply seem to offer a potential remedy for the problem of transporting medical knowledge to the location where it is needed most.

Primary medical care in developing countries is often provided by personnel with a very heterogeneous level of medical training, commonly called village health workers. The main tasks of a village health worker are: to provide a diagnostic classification of a patient's disorder as an easily treatable and not life-threatening disease or as a potentially dangerous disease; if appropriate, to deliver a treatment scheme compatible with the health worker's proficiency and therapeutic resources; to refer patients with potentially dangerous diseases to a professional nurse or doctor.

The availability of professional medical care has a direct bearing on the extent of a health worker's activities. If information exchange with properly trained medical personnel is available, the quality of medical care provided by health workers is improved. Measures that should be taken to maintain an adequate standard of primary medical care, therefore, include implementing policies to provide for on-the-job training to improve efficiency and effectiveness and providing reference materials that are easy to consult in unclear cases.

The purposes of the project discussed here are: to establish a knowledge-based decision support system for village health workers covering the most common diseases in developing countries; to implement this knowledge-based system on portable microcomputers; to include tutorial features as well as a reference structure in the design of the system to meet the demands of health workers on the job.

In the following we present a description of the system specification and design, show the structure of knowledge representation and system implementation, and, in the Appendix, give sample dialogues between a health worker and the decision support system.

System Design

Ample experience with the principles of teaching medical care to health workers has been accumulated through the years (Werner 1980; Werner and Bower 1982; Abbatt and McMahon 1985). In the design of our system we strove to draw from that vast body of experience as much as possible to avoid mistakes frequently encountered in intercultural exchange programs. For that purpose, the standard textbook *Where There Is No Doctor* by Werner (1980) served as the primary source of reference. In addition, medical knowledge was extracted from the relevant literature (King et al. 1978, 1979; Mason-Bahr and Apted 1982; Upunda et al. 1983).

As a second source we used the experience gained by G. Porenta, who, as a member of the Austrian Committee for Ethiopia, worked for 2 months in an ambulatory care center in the Tigre province of northern Ethiopia. Working together with village health workers of different educational levels, he experienced the need for a flexible source of information to cope with problems arising in primary medical care. Also, we used health statistics (see Table 12.1) showing the actual distribution of diseases within a 3-week period as a guideline for the selection of diseases in the design phrase.

The data given in Table 12.1 show that a small group of diseases accounts for a large percentage of the disorders encountered. Accordingly, treatments of diarrhea, worn infestations, eye diseases, skin afflictions, and malaria made up approximately half of the daily work. Concentrating on providing good quality health care in these areas would significantly improve the overall performance.

The disease distribution presented pertains only to a specific situation that includes the geographic area (highland), political context (civil war), social environment (poverty, famine), infrastructure of health care (rural health centers closed, two town hosptals), and other factors. It is not possible to devise a detailed generic general-purpose knowledge-based system suited for village health workers in different countries. For proper function and acceptance, the system has to be tailored both to the region where it is used

FIGURE 12.1 Disease Profile of Northern Ethiopia*

Disease	Cases	%	Sum %
Diarrhea	386	14.8	14.8
Worms	371	14.2	29.0
Eye disease	297	11.4	40.4
Skin disease	203	7.8	48.2
Infectious diseases			
Malaria	157	6.0	54.2
Bronchitis	146	5.6	59.8
Ear infection	102	3.9	63.7
Urinary tract	75	2.9	66.6
Throat	57	2.2	68.8
Pneumonia	50	1.9	70.7
Tuberculosis	46	1.8	72.5
Bilharzia	3	0.1	72.6
Others	714	27.4	100.0

*Over a 3-week period diseases as encountered in an ambulatory care center were recorded.

and to the people using it. Therefore, the system presented in this study is a specific instance that applies to the northern part of Ethiopia, although general prnciples as outlined in the book by Werner (1980) are included.

In designing the system, we identified five problem areas to be represented (diarrhea, infestation with worms, eye diseases, skin diseases, common infectious disease) and three different entry points into the system (diagnosis, therapy, drug prescription).

If decision support for diagnosis is requested by the user, the system starts with a question-answer strategy to find an appropriate diagnosis, to suggest a treatment plan, and to give detailed advice about drug prescription if necessary. At this level we assume that the village health worker is familiar with the most common medical terms so that he or she can answer each question with "yes," "no," or "unknown."

The diagnostic strategy is structured around the five kernel areas of disease mentioned above. In the first step, the system tries to ascertain that the medical problem as presented by the patient is not dangerous and therefore falls within the range of competence of the village health worker. If none of the danger criteria is met, the system then goes on to ask questions about diseases ranked by their frequency of occurrence in an ambulatory care center. If danger criteria can be ascertained, the system suggests that the patient be referred to a doctor or a professional nurse.

After deciding on the main complaint, the village health worker is prompted to answer questions posed by the system. A choice of entering a set of symptoms in free order is not offered. With this strategy, we attempted to elucidate the diagnostic pathways of the system with a tutorial perspective. By following a predetermined sequence of questions, the village health worker should become acquainted with a standardized way of efficiently taking the history and performing a very coarse physical examination.

In the present stage, questions are asked by displaying text on the screen using the terminology of Werner (1980) (for examples, see the Appendix). Clearly, this is not the most appropriate form of a user-friendly interface. An icon-based dialoque using symbols instead of text, with a pointing device for the user input, might be more appropriate for a broader audience and might also feature language independence (Trappl and Horn 1983). In developing our prototype, however, we concentrated on the proper representation of the knowledge and set aside the problem of an adequate user interface.

At the second step, the system uses the diagnosis established in the first step to suggest an adequate treatment plan consisting of general advice and drug prescription if deemed necessary. It also provides the opportunity to start with a diagnosis independent of a pass through the diagnostic workup. While working on a treatment plan, questions about contraindications and, if not yet known, about general patient data (e.g., age, weight) must be answered. Finally, the treatment of choice is presented.

While determining an adequate treatment scheme, the system evaluates its knowledge about contraindications, side effects, treatments of choice, and price information pertaining to the drugs in its knowledge base. In the current implementation, the treatment scheme associated with the lowest cost for a complete course of treatment is selected from the set of equally effective schemes. Therefore, if the system is used for treatment selection on a regular basis, a significant cost reduction in drug expenditure will follow. User access to this module of the system is available independently of the diagnostic or therapeutic branch.

Implementation

A prototype of the system has been implemented on a portable personal computer (PC) by means of VIE-KET, the Vienna Knowledge Engineering Tool (Pfahringer and Holzbaur 1985). Similar to KEE (Kehler and Clemenson 1983) and BABYLON (Primino and Brewka 1984) VIE-KET is a hybrid knowledge engineering tool that offers subsystems for handling various knowledge representation schemes such as frames, rules, Prolog, and LISP. In principle, the system could have been developed in a purely procedural language like Pascal, BASIC, or FORTRAN. However, the developing environment associated with a hybrid knowledge engineering tool significantly improves and simplifies the development process.

According to the specifications worked out during the design phase, the system consists of several modules assigned to deal with specific domains of knowledge. As the corresponding problem areas differ in their specific structure, the modules of the problem differ in their implementation in VIE-KET.

Modules for Diagnosis

In the two problem areas "diarrhea" and "infestation with worms," expanded decision networks guide the health worker to the diagnosis. At each node, a node interpreter evaluates and weighs a set of symptoms represented as premises of a rule and branches according to the relation between a threshold value and the sum of weights. In the simplest case, the premises of the rule are classified into IS-PRESENT (IP) and MAYBE-PRESENT (MP). Finding one IP premise or two MP premises causes continuation on the YES branch of the decision node. Figure 12.1 shows as an example the decision network covering the problem area diarrhea.

Within the system, the expanded decision networks are represented using a frame structure. A general frame NODE with slots IS-PRESENT, MAYBE-PRESENT, YES, and NO is defined. Attached to each NODE frame is method NEXT-NODE that determines which branch will be followed after evaluation of the premises. Networks then can be built incrementally by creating new instances of the general frame NODE. Advantages of choosing frames as the representation tool include flexibility, a simple editing procedure, and graphic display routines for debugging purposes. During an actual

FIGURE 12.2 Decision Network for the Problem Area "Diarrhea." [If one of the IS-PRESENT symptoms (IP) is confirmed the yes branch (y) is pursued.]

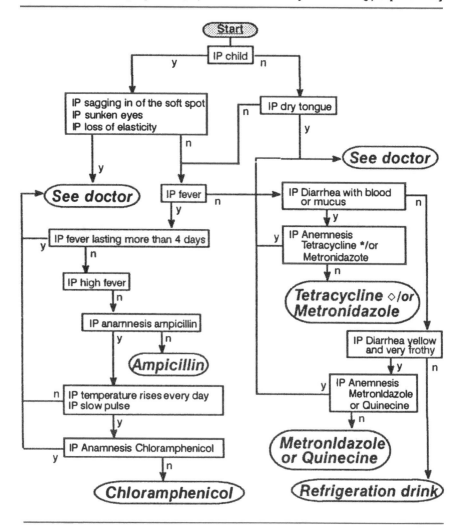

consultation session, the tree is traversed by sending NEXT-NODE messages to the actual node and updating the node accordingly until the proper form of diagnosis is reached.

For the diagnosis of diseases affecting the skin and the eyes, a rule-based approach is more adequate. Each symptom is associated with two certainty factors ranging from –1 to + 1. The first factor characterizes the relation between a symptom and the disease currently in focus, where + 1 indicates a pathognomonic symptom and –1 indicates exclusion of the disease if the symptom is present. The second factor gives an indication of the relation between the absence of a symptom and the disease. Figure 12.2 shows the premises of two rules concluding in the diseases "bed sores" and "bad circulation" under the sub-heading "large open sores or skin ulcer."

For illustration, Figure 12.3 exemplifies the two relationships in the case of the disease "herpes" and the symptom "location on one side of the body." If manifestations of the disease are present strictly unilaterally, supportive evidence with a strength of 0.4 is concluded toward the diagnosis of herpes. If symptoms are present bilaterally, evidence of 0.7 against a diagnosis of herpes will be processed.

In selecting the certainty factors, we opted to restrict the positive and negative values, respectively, to 1, 0.7, 0.4, 0.2, and 0. Two physicians then independently scored the symptoms in relation to the diseases. This scoring scheme was conducted according to the physicians' personal beliefs, and in

FIGURE 12.3 Schematic of Two Rules in the Problem Area "Skin Disease" for the Subtopic "Large Open Sore of Skin Ulcer."

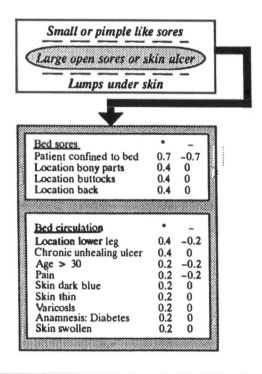

FIGURE 12.4 Relationships Between the Symptom "On One Side of the Body" and the Disease "Herpes."

Herpes zoster

Symptoms	+	–
■■■		
Location on one side of the body	0.4	–0.7
■■■		

most cases the certainty factors turned out to be identical for the two physicians. In diverging cases, discussions led to a consensus.

It should be pointed out that the certainty factors assigned to each symptom do not depend only on the relationship between a single symptom and the corresponding disease. The set of diseases that are potential candidates for a diagnosis decreases as the diagnostic process proceeds. As diseases are worked on sequentially by the system, knowledge about which diseases have already been rejected by the system is available. In scoring the symptoms of diseases that appear late in the diagnostic pathway, this narrowed solution space has a direct bearing on the importance and weight of certainty factors. The rank order of diseases therefore influences the associated certainty factors.

An algorithm similar to the formula in the MYCIN system is used to combine certainty factors for the different symptoms to arrive at a final score. If this score exceeds 0.8, the diagnosis of the disease presently evaluated is concluded and the diagnostic procedure stops.

During the diagnostic process, the age and weight of an patient must be provided. For all children under age 12, a malnutrition module then checks their nutritional status according to a simple algorithm (weight for age) commonly called "road to health" (Werner 1980). As malnutrition poses a major threat to the well-being of infants in developing countries, the system points out to the health worker the computed nutritional status of the child is in units of percentage of normal. He or she is then expected to take appropriate actions.

Modules for Therapy

For the drug module, knowledge about usage and dosage of drugs is represented in a frame structure that can combine both declarative and procedural knowledge. Declarative knowledge about medication includes the period of time a drug should be given to reach an optimal effect, the contraindications that have to be checked before administering a certain drug, warnings about side effects, and additional advice related to drug usage. Procedural knowledge is applied to compute the required dosage for a specific patient. The algorithm then suggests the number of tablets or capsules that should be taken every day for a certain period of time according to the prescription stored in the knowledge base.

The frame hierarchy includes an inheritance mechanism that makes it possible conveniently to represent different drug dosages for different diseases without creating two entirely different structures. For example, the frame "mebendazole" holds knowledge about this anthemintic drug including default dosage, contraindictions, side effects, and warnings. For the special case of infestation with threadworms, the default dosage, however, is not appropriate. Therefore, another instance of the frame "mebendazole" is created containing only the special dosage for this disease. The rest of the information is inherited from the original frame.

Depending upon the availability of drugs, the system could arrive at several different drugs as the treatment of choice. In this case price information is used to select the cheapest treatment scheme to be suggested to the health worker. The differences between two treatment plans can be clearly demonstrated in the treatment of ascariasis. Administering mebendazole is on the average 7.5 times more expensive than treating with piperazine. Besides drug prescriptions, the system also suggests other therapeutic actions, if appropriate, such as bed rest, diet, or inhalations.

Discussion

The present system provides for the transfer of simple medical knowledge. Without field tests, however, it is difficult to assess how adequately the chosen structure serves this purpose. Clearly, several shortcomings deserve further discussion.

Proper use of this system requires the user to be familiar with common medical terms. The system is not designed to interact with users who might need explanations of the medical terms encountered in the dialogues. A smart dictionary could be developed to accomplish this task.

The development of a language-independent user interface relying on image-based computer interaction would provide an interesting and useful extension of this system. However, most of the currently available computer systems that are also truly portable lack the technical capacity to allow an implementation using high-resolution graphics.

The system uses question-answering strategy to guide the user through the diagnostic and therapeutic workup. This rigid and structured approach still awaits evaluation. It could well be that a flexible strategy with free entry of signs and symptoms and subsequent diagnostic and therapeutic workup will turn out to be more effective.

Several other attempts have also been developed to provide decision support for health care in developing countries on various levels. TROPICAID represents a similar approach focusing on a microcomputer implementation. This diagnostic system is a joint project of Medecins Sans Frontieres and the Universite Pitie Salpetriere, Paris, designed for use in the Tchad (Auvert et al. 1986). Field tests of that system are well under way and additional projects are in a planning phase. A preliminary evaluation provided evidence that a microcomputer-based system can support the daily work of medical personnel in a useful way.

During the developmental phase, the ease of handling and the flexibility of an open system allowing access to the programming language proved to be important. Expert system shells usually abide by the paradigm of production rules and cannot easily support two certainty factors for one premise. In this case a problem sometimes has to be adjusted to the developing tool, whereas with open systems the tool can be tailored to the problem. As a trade-off, programming effort is increased when using open systems.

Implementing a knowledge-based system on a personal computer naturally imposes restrictions on the design because of the limited resources. To date, commercially available tools for PC applications have lacked either sophistication or efficiency or even both. Maintenance of even a small knowledge base is by no means trival, especially if more than one expert contributes of the knowledge. Usually, a choice between two alternative approaches must be made: either allotting significant resources to the construction and maintenance of an efficient and fast knowledge-based system because insufficient tools are available on PCs (thus sacrificing the rapid prototyping paradigm) or using the available tools for prototyping, thus forcing the user to spend considerable time and attention while using the system (not an option with the intended audience of the present project).

For future projects, a reasonable trade-off would include the development of the system on a dedicated workstation for rapid prototyping and on its completion, transfer to a custom PC for delivery of the final application using appropriate run-time versions. However, the software garden as yet is only sparsely populated with suitable tools.

Conclusion

In summary, for the knowledge-based system presented here several aspects must be worked out in more detail before field tests are eventually considered. The main problem, still unsolved, is the design of a user-friendly interface tailored to the educational level of village health workers. In addition, a disease profile for the region of intended use must be established to restructure the knowledge base for that specific application. However, from our experience with this system we conclude that knowledge-based systems can be a valuable contribution to the international effort to bridge the gap in the quality of health care between developing and industrialized countries.

Appendix

The following two dialogues exemplify the course of an interaction between the system and a health worker. In the dialogues, user inputs are marked with **. Comments on the dialogues appear in parentheses.

(The first consultation session deals with a patient suffering from small open sores with itching especially at night. Most of the lesions are covered with yellow crusts. A preferred location seems to be the spaces between fingers.)

```
What is your problem:
   ** Skin **
      Eye
      Worms
      Diarrhea
      Fever
      Cough
      Urine reddish
      Painful urination
```

(After the problem area "skin" has been selected, another menu is presented:)

```
What is your major skin problem:
   ** Small or pimple-like sores **
      Open sores or skin ulcer
      Lumps under the skin
      Large spots or patches
      Warts
      Rings (spots with raised or red edges)
      Blisters
```

(After the heading "small or pimple-like sores" has been chosen, the system generates a hypothesis about the disease and tries to test it by asking about additional relevant symptoms:)

```
      Do you have the following symptom:
         itchiness/scratching? ** Yes
      Do you have the following symptom:
         itchiness at night? ** Yes
      Are the symptoms located on the:
         head? **?
```

(A question mark as an answer to the preceding question prompts the system to display information about the disease currently in focus:)

```
PRESENT p    ABSENT a    UNKNOWN u   NOT1-ASKED N
Scabies                  Believe: 640
        itchiness/scratching              400   -400   p
        itchiness at night                400      0   p
##      loc – head                       -700    200   n
        loc – face                       -700    200   n
        Combination:                      600   -200   n
          loc – genitals                  400      0   n
          loc – between fingers           400      0   n
          loc – body                      400      0   n
```

("Scabies" is the disease hypothesis presently active. "Itchiness/scratch-ing" and "itchiness at night" are known as being PRESENT and are flagged with a "p." Both of these symptoms have a certainty factor of 400 associated with them, yielding a combined value of 640 ("Believe: 640"). The symptom "loc head" is presently being asked ("##") and like all the other symptoms is designated as not yet asked ("n"). The system repeats the question:)

```
Are the syptoms located on the:
  head? *No
Are the symptoms located on the:
  face? **No
Are the symptoms located on the:
  genitals? **No
Are the symptoms located on the:
  between fingers? **Yes
```

(At this time, the diagnosis "scabies" is confirmed.)

```
*Diagnosis*: Scabies
Do you have the following symptom:
  yellow crusts? **?
```

(Now questions concerning the treatment scheme are asked and explained on prompting the system with "?":)

```
yellow crusts is a DANGER SIGN of Scabies
```

(Yellow crusts indicate a superinfection, which in turn would require some kind of drug treatment.)

```
Do you have the following symptom:
  yellow crusts? **Yes
```

(Penicillin is one of the drugs, that can be used. So contraindictions of penicillin will be checked:)

```
Do you have the following symptom:
  Penicillin allergy? **Yes
```

As penicillin cannot be used, sulfadimidine is considered as an alterna-tive treatment plan:)

```
Do you have the following symptom:
  Sulfonamide allergy? **No
Enter AGE in years (in case of babies you can use months as a
  unit, e.g.: 4M(onths)) Age:   *23
```

(As babies should not be treated with sulfadimidine, the patient's age was requested. The next question concerns pregnancy:)

Is the patient
female? **Yes
Is she
pregnant? **No
Do you have the following symptom:
dehydration? **No

(As the patient is not dehydrated either, sulfadinidine can be used. To compute the dosage, the weight or the age of the patient is used:)

Enter WEIGHT ing (or U if unknown): **U

(No weight was given, but as the age is already known, a therapy is now presented:)

Therapy

Scabies

* Everyone in the family should be treated.

* Bathe and change clothes daily.

* Ointment Lindane/Vaseline:
Lindane 1 part (of 15% solution)
vaseline (or body oil) 15 parts:

Wash the whole body with soap and hot water;
heat the Vaseline and the Lindane and stir well;
smear the ointment on the whole body
except for the face,
leave for one day and then bathe well;
After treatment, put on clean clothes and
use clean bedding.
Repeat this treatment one week later.

* WARNING: Can cause poisoning if used too often.

(To treat the superinfection an appropriate dosage for sulfadimidine and warnings are displayed:)

Sulfadimidine: 1000mg, 4 time(s) a day
Warnings (Side effects):
Take it with lots of water
If there is rash or itching after taking the tablets, then
stop taking them and drink lots of water!
May cause: vomiting, diarrhea.

(Now the first sample consultation is finished. The second case deals with the problem area "diarrhea":)

What is your problem:
    Skin
    Eye
    Worms
 ** Diarrhea **
    Fever
    Cough
    Urine reddish
    Painful urination

(For this consultation session, the patient has the problem of yellow diarrhea without fever.)

> Enter AGE in years (in case of babies you can use months as
> a unit, e.g.: 4M(months)) Age: **15
> Do you have the following symptom:
> dry tongue? **No

(Dehydration is the main concern when treating diarrhea. Therefore, it is necessary to look carefully for signs of dehydration. For adult patients, checking the appearance of the tongue might serve that purpose.)

> Do you have the following symptom:
> fever? **No
> Do you have the following symptom:
> Diarrhea with blood or mucus? **No
> Do you have the following symptom:
> Diarrhea yellow and very frothy? **Yes

Diagnosis has again been established. Before suggesting a proper treatment plan, the system makes sure that this treatment has not been given recently without success).

> Do you have the following symptom:
> Anamnesis: metronidazole or quinacrine? **No

(And now the diagnosis:)

> *Diagnosis*: giardia

(Currently considering metronidazole:)

> Is the patient female? **No

(As the patient is male, no question concerning pregnancy is asked. So metronidazole will be suggested:)

> Therapy
> Metronidazole: 250mg, 3 time(s), for 5 days
> Warnings (Side effects):
> Do not drink any alcohol.
> May cause: nausea, headache.

References

Abbatt, F., and R. McMahon. 1985. *Teaching Health-Care Workers: A Practical Guide.* London: Macmillan Education Ltd.

Auvert, B., P. Aegerter, V. Gilbos, E. Benillouhe, P. Boutin, G. Desve, M.F. Landre, and D. Bos. 1986. Has the time come for a medical expert system to go down in the bullring: The TROPICAID experiment. In *Cybernetics and Systems '86,* ed. R. Trappl. Dordrcht: Reidel. 727–734.

Kehler, T.P., and G.D. Clemsenson. 1983. *KEE—the Knowledge Engineering Environment for Industry.* Palo Alto, CA: Intelligenetics.

King, M., F. King, and S. Martodipoero. 1978. *Primary Child Care, Book I, A Manual for Health Workers.* London: Oxford Univ. Press.

____. 1979. *Primary Child Care, Book 2, A Guide for the Community Leader, Manager, and Teacher.* London: Oxford Univ. Press.

Manson-Bahr, P.E.C., and F.I.C. Apted. 1982. *Manson's Tropical Diseases.* 18th ed. London: Basillsere, Tindall.

Pfahringer, B., and C. Holzbaur. 1985. VIE-KET: Frames + Prolog. In *Oesterreichishe Artificial Intelligence—Tagung,* ed. H. Trost and J. Retti. Berlin: Springer. 132–139.

Primino, F. di, and G. Brewka. 1984. BABYLON: Kerrasystem einer integrieden Umgebung fuer Entwicklung und Betriebvon Expertensysteanen, Forschungsbericht der GMD, Bonn.

Trappl, R., and W. Horn. 1983. Making interaction with a medical expert system easy. In *MEDINFO 83,* ed. J.H. van Benunel, M.J. Bail, and O. Wigertz. Amsterdam: North-Holland. 573–574.

Upunda, G., J. Yudkin, and G.V. Grown. 1983. *Guidelines to Drug Usage.* London: Macmillan.

Werner, D. 1980. *Where There is No Doctor.* London: Macmillan.

Werner, D., and B. Bower. 1982. *Helping Health Workers Learn.* Pal Alto, CA: Hesperian Foundation.

13

Financing Scheme Portfolio for Health Services in Developing Countries: An Expert System Approach

Eckhard F. Kleinau

Introduction

Governments of developing nations are faced with increasing costs of health services at a time when public spending becomes more and more constrained. This is leading to a change from a free care policy to cost sharing with increasing patient contributions. Many different ways exist to finance health services beside direct payments for care. Not all possible financing schemes are considered by policy makers, partly because their revenue potential and effects on the accessibility to care by less affluent populations are ill understood. An expert system (ES) can assist in making choices in this situation of insufficient information, considerable uncertainty and complex interaction among socioeconomic, health related and other factors.

The complexity of implementing health care financing schemes is suggested by the following illustrative list of parameters to be considered simultaneously:

- Recurrent costs of services offered;
- Historic utilization rates and future development;

The author gratefully acknowledges the support and encouragement by Jerry Mechling from the Kennedy School of Government in writing this paper and Charles Mann with regard to its further development for this book. The knowledge presented here also profited from extensive research in the field of health care financing together with Donald S. Shepard at the Harvard Institute for International Development and the Harvard School of Public Health. I would like to thank the fellows of the Takemi International Program at the Harvard School of Public Health, class 1990, for their valuable comments and critique.

- Demand for different types of health services and their price and income elasticity;
- Disease incidence and need for services (standard);
- Demographic data;
- Quality of care offered;
- Ability and willingness to pay of the target population;
- Availability of subsidies and other sources of financing;
- Factors influencing the implementation: politics, socioeconomic environment, technical support, health service infrastructure, management information and control systems, etc.

The issue is further complicated if one wants more elaborate systems and gives consideration to equity and accessibility, which include:

- Price discrimination by type of service;
- Price discrimination by the ability to pay;
- Price discrimination by patient distance;
- A health insurance scheme with risk sharing;
- Demand and price elasticity for vital, essential and non-essential services or interventions (i.e. treatment of acute malaria, hernia repair, general consultation).

It is difficult for health service and program planners to consider all the information, select a sustainable cost recovery scheme, and set an appropriate price level. The purpose of this paper is the development of a Financing Scheme Portfolio Expert System which, for a certain scenario, proposes the best possible mix of schemes and prices to decision-makers. An experienced, human expert will be able to deal with this complex situation. However, few government and local or regional offices of development agencies dispose of sufficient human resources.

This paper describes the potential user group within government and agencies providing technical support and their potential willingness or resistance to use such an ES. Some thoughts are given to the approach of successfully implementing this ES in an environment where the number of problem solving and decision facilitating software increases. Features of the ES which could be key success factors with regard to acceptance are pointed out.

Schemes for Financing Health Care in Developing Countries: Health Care Financing Schemes

Central government spending in the health sector has been falling in many developing countries between 1972 and 1985 (World Bank 1988). It amounts to as little as 2% in several countries of Sub-Saharan Africa. Due to high inflation rates and a rapidly growing population the per capita expenditure for health care fell much more drastically, close to 50% in Senegal and Ivory Coast between 1980 and 1985 (Vogel 1988).

In most developing countries government services include free or low cost curative and preventive health care, specialized disease control programs, immunization, sanitation, clean water supply, and others. To fulfil its commitment the government budget for these activities would have to be increased substantially. Governments allocate available funds to more visible activities such as curative care, amounting to 70 – 90% of total expenditure for health care. Other activities which are much more cost-effective in terms of costs per additional life saved such as vaccinations receive relative little financial support. However, even services receiving the greatest share are severely underfinanced, unable to guarantee services at an appropriate level. Tertiary level care in referral hospitals is generally much more subsidized than primary care. This leads to an inappropriately high utilization of higher levels of services for diseases that could be treated more cheaply on lower levels.

Great discrepancies in financing exist for similar health services between rural and urban areas. This implies that low income areas receive a lower than proportionate share of public expenditure. This is contrary to the principle of equity proclaimed by many government programs. The fact that limited supplies are often provided free for certain groups such as civil servants, but that they are insufficient for less affluent populations further widens the gap. Higher income households in urban areas in some Asian countries profit from five times higher health sector subsidies than rural low income households.

The problem of underfinancing of public sector health services leads governments more and more to search for additional sources of revenue. Some have started charging for services. However, due to the sorts of social and political implications mentioned above, initiatives have been hesitant and half-hearted. Rwanda, for example, covers with fees for service only 7% of its health sector recurrent costs (Shepard, Carrin, and Nyandagazi 1987; Shepard and Nyandagazi 1988). The uncertainty is increased by limited knowledge about possible effects of the different methods of health care financing. Few of the possible means are employed at all and it seems that decisions are not preceded by a formal evaluation of all feasible schemes. The reasons for this shortcoming will be further investigated in sections describing the decision-making process and problems.

Private health institutions in countries with a free care policy for a long time have been charging for their services in one form or another. Most not-for-profit institutions like missions cover 60 – 80% of their recurrent costs through direct patient contributions. While mission facilities are more frequently operated in low income areas, private for-profit care providers are mainly found in urban settings. The lesson that government can learn from these not-for-profit and for-profit providers is that government has underestimated the potential and willingness of the population to pay for quality care.

Table 13.1 lists several health care financing schemes presently in use. Their effectiveness and range of application varies greatly. Many have been tried on a very small scale, for example, community labour. Others are implemented on a national level for example, fees for service.

TABLE 13.1 Health Care Financing Schemes

GOVERNMENT BUDGET ASSESSMENT, STRUCTURAL ADJUSTMENT, REALLOCATION
- Shift priorities: curative-- > preventive
- More effective structures: large-- > small units
- Ownership: public-- > private
- Management & administrative capacity development

SHIFT RESPONSIBILITY TO MORE AFFLUENT & EFFICIENT SECTORS
- Nutrition-- > Agriculture
- Hygiene-- > Water & Sanitation & Housing (urban development)
- Health education-- > Education

NEW TAXES (revenue committed, social insurance)

INSURANCE DEVELOPMENT
- Decentralized & local pre-payment systems (voluntarily)
- Cooperative & employer based
- Local social insurance (mandatory)
- Regional & national social insurance

COMMUNITY PARTICIPATION
- Fee for service, not-for-profit, revenue within community
- Drug sales
- Pre-payment, revenue managed by community
- Production-based prepayment
- Income generating schemes, intersectoral
- Community labour
- Individual labour & donations
- Fund raising: donations, fines, festivals
  - support from family members employed in cities, abroad
- Community activities, campaigns
- Community taxes (non-government, traditional system)

PRIVATE SECTOR
- NGOs & PVOs
- Fee for service, drug sales (both for profit and not for profit)
- Pre-payment by individual & household
- Private philanthropy
  - gifts, donations by individuals or groups
  - domestic or international
- Private saving clubs

Decision-Making Process
in Health Care Financing

This paper focuses on two different decision-makers concerned with the development and implementation of health care financing schemes. For each of these the process of decision-making will be analyzed according to the framework suggested by Gerrity (Keen and Morton 1978): *Intelligence, Design, Choice, Implementation and Control.* During the development of the Financing Scheme Portfolio ES this process will have to be studied formally prior to the creation of an appropriate computer model.

Decision-Making in a Ministry of Health

The first type of decision-maker is found on director level of national Ministries of Health, such as the directors for planning, hospital services and preventive medicine. Even though the planning section is more directly concerned with investment rather than recurrent costs each development activity requires finances for operating and maintaining the facilities later on. The preventive medicine section is usually the most constrained.

Decisions, however, are not implemented on this level. They are forwarded in the form of recommendations to the Ministry of Finance and even the presidency or parliament. The discussion that follows makes the optimistic assumption that these recommendations do in fact influence decisions of politicians, at least to some extent.

Decisions about financing health services are essentially required on three occasions: for the development of health sector activities for a five year plan, to prepare the annual budget proposal of the Ministry of Health, and to propose long-term financing of special programs. A major decision could be whether to allow sales of drugs at low cost at government facilities rather than providing drugs free. The process of decision-making can be described generally as follows (considerable differences in detail exist between countries):

Intelligence. Each service within the health sector has a list of facilities, programs and activities which are carried out on each level. The budget is itemized according to major categories, such as fuel, maintenance, drugs, personnel etc. Each facility or program might have its own budget or they might be combined into the budget for an administrative region. Personnel and drugs are handled in different ways. In a centralized system they might not be included in the facility's or program's budget. Personnel might be paid by a different ministry (like Public Services). Drugs might be purchased nationally and distributed to regions.

The real costs to provide adequate services on all levels are rarely calculated. Individual units might or might not be requested to formulate their proper budgets, with no guarantee that this will influence central decision-making. For programs and services supported by foreign donors such information is usually available.

Three pieces of information seem to be the most important for the budgeting process: last year's allocation, cost increments and inflation, and some a *priori* knowledge about the amount available for the coming period.

Design. Last period's expenditures are used to estimate those for the following period adjusted upwards for inflation and other cost increments. If that turns out to be considerably more than the anticipated funds, budget items are reduced proportionally. Usually no consideration is given to any gap which might exist between amounts budgeted and actual costs to provide quality service. Performance based resource allocation is not employed. There is little incentive to develop alternative financing schemes, unless they are government policy.

Often governments are persuaded by foreign donors to generate additional revenues through fees for service, for example. In many cases this is just a token rather than a serious attempt to finance services. Such an attempt becomes even more futile when a large proportion of patients is exempted, no control is exerted on the collection process, and revenues have to be forwarded to the Ministry of Finance rather than augmenting the health unit's budget.

Government programs supported by donor agencies are likely to have their budgets based on planed operations. Rather than listing all operations along with their financial sources, frequently the health sector budget is restricted to activities requiring direct government financing, omitting foreign contributions. Because donor support is given for a limited time only, such programs eventually face the same financial problems as regular government health services.

Choice. As explained above there is not much of a choice between different financing schemes at the ministerial level. The ministry will combine each service estimate into a single budget proposal, usually requiring services to cut back. Who looses or who wins in this game depends on each player's political influence and power. Curative and hospital services have a stronger position than others. They are less likely to give up resources. Income from fees for service cannot be considered as a revenue to an institution if it operates like a special government tax on health care and does not benefit the health sector directly.

On the national level the Ministry of Health has to defend its budget against other competing ministries. Its position is not strong. Education, agriculture or defence are usually given higher priority. A cap on overall spending for health leads to further reduction of health service budgets.

Implementation. Once the national budget for all ministries is approved the actual allocation of funds frequently takes place through the Ministry of Finance on request of the Ministry of Health. Every country has its standard operating procedures. Even though the approved amount should be expended during a financial year, restrictions not made explicit could further curtail funding. There is no incentive for health institutions to enforce fee collection, unless they are allowed to add this revenue to their operating budget.

Control. Expenditures are recorded according to standard accounting procedures at the final recipient of funds and checked on a local basis. The central administration receives periodical, mostly annual, financial or cash flow statements. These statements contain usually all sources of revenue, government subsidies, fees collected, and outside contributions. Detailed control occurs only in case of severe irregularities.

Decision-Making in Donor Agencies

The second type of decision-makers concerning health care financing are program officers in regional offices of international or bilateral agencies (WHO, World Bank, USAID, GTZ, etc.). The process to develop a financial framework is substantially different from that in national ministries. While limited government resources are spread more or less evenly over all existing institutions leading to underfinancing of many of them, donor funds are concentrated on selective activities and programs. These can be considered fully financed according to the plan of operations. However, donor funds are limited as well as government ones, leading to geographically and service level restricted interventions. The problem of long run financial sustainability after donor support has come to an end is given too little attention when planning projects.

Intelligence. Donor supported projects generally have good information about the demographic, epidemiologic, socioeconomic, and administrative situation for their area of intervention. Such information is available from surveys and studies carried out by national and international institutions or by the donor agency itself prior to project design and implementation. A plan of operations quantifies targets for the project duration. Potential sources of revenues apart from the donor's contribution are identified. Population based utilization rates, staffing norms and unit costs are calculated and projected. However, despite detailed information demand for care is difficult to derive.

Design. Budgets are itemized, usually separately for different programs, in a similar form to those of the Ministry of Health. They are frequently established in collaboration. Because the aim is a complete coverage of recurrent costs and long-run sustainability of program activities, consideration of alternative sources has become mandatory. Frequently the following sources are evaluated: government subsidies, community contributions and sales of drugs at low cost. Other possible schemes are often ignored. Insufficient funding leads to a cut-back in program activities rather than in underfinancing.

The evaluation of financing schemes is further complicated because their suitability depends on the type of services. Again, for curative care there are several options with a good potential for cost recovery. For example, drug sales or user fees. Preventive care like family planning or prenatal care are likely to generate little funds when using the same methods. Moreover, demand might drop considerably with charging for such services. On the contrary price elasticity of demand for curative care is much more inelastic.

The potential of revenue generation by those sources included in a project design seems often underestimated (ignoring the fact that many projects might be not affordable by the host countries). Otherwise it is hard to explain why in projects donor contributions range from 50 to 100% of the recurrent costs (Bossert et al. 1987).

Choice. The choice is usually not for a single best financing scheme but rather for an optimal mix of different schemes. Some projects are able to estimate an expected amount of revenue for each source. Mostly, however, the principle of trial and error is followed. Some schemes are simply included because it is the policy of the donor agency, not because of their potential revenue generation. For example, community contributions. Other concerns, such as affordability and equity, influence the selected strategy as well.

Within a donor agency projects are competing for funding in a similar manner as ministries. What gets funded and at what level often depends on personal relations within the organization and with the outside (congress, financing ministries, etc.). In general one can assume that projects which explicitly state and quantify host country financial sources receive today more positive attention than those which do not.

Implementation. The implementation of financing schemes can prove difficult due to managerial problems or low levels of acceptability on the side of the host country and the target population. These factors can be underestimated despite elaborate project preparation and detailed plans of operations.

Control. The control of donor funds is usually very tight and accounting principles to be followed are rigid. This is mostly not true for other sources. Government subsidies and community contributions often remain below target.

Management Level and Problem Structure
(Applying the Anthony x Simon Matrix)

The Anthony x Simon matrix shows the relationship between three levels of management (operations control, management control, strategic planning) and the decision-making structure in a three by three matrix. Decision problems range from very structured on operations control level, to semi-structured on management control level, and to unstructured when strategic planning is involved (Gorry and Morton 1971:13(1)). Based on the location of an actual problem in the matrix different demands are exerted on data quality and information technology. Less structured problems with strategic importance – like health care financing, for example – require a decision support or expert system solution. Structured problems on operations control level are solved with office automation or management information systems.

The level of personnel in ministries and donor agencies involved in the decision-making process varies according to the stages within the analytic framework described in the previous section. Intelligence gathering and aggregation of data is and can be carried out by less experienced personnel. Senior analysts will be involved in the development of health care financing schemes and alternatives, the design phase. Tasks on this level are less well defined and require broader knowledge about socioeconomic factors or the effect of prices in relation to services offered. Directors or senior program officers are mainly involved in the choice phase. Factors hard or impossible to quantify have to be taken into account. The better the previous two phases have been prepared the more effective will be the third.

The problem is partly structured. Costs of services can usually be calculated based on available data. Revenues from some sources, like government subsidies or sale of drugs in urban areas, can be estimated with reasonable precision. Demand for care or price elasticity has been studied in a few countries and restricted settings. Expertise and judgement is needed to estimate their effect on cost recovery. The problem becomes unstructured when political and economic factors are included. In general the development of health care financing schemes can be considered complex and semi-structured to unstructured in the Anthony & Simon matrix. It is located somewhere between the management control and strategic planning level. Long-run financing of recurrent costs has a strong strategic component, not only for health services in developing countries.

This would imply less stringent requirements for information accuracy and frequency. Data sources can be soft, in fact many data are derived from small studies which build their validity on a number of assumptions. Few data originate from large-scale data processing, for example from national surveys or from large hospitals in developing countries. The level of aggregation of available data has to be high. Even though one is concerned with the present and future status of health care financing the recommendations of the expert system will draw on past experience. This would suggest also the possibility of testing the ES by tracking with historical data to test the performance of the ES.

Problem Description

The reason that only few financing schemes are employed in the health sector is not only political inertia. The problem involves a multitude of parameters, complex interactions, and uncertain effects. Many factors have to be considered to estimate the potential effect of each. Questions like the following have to be answered: "What will happen to demand for care if one charges for services, given a certain family income?"; "How will utilization change if people are insured rather than paying out of pocket?"; "How will the demand for preventive care be affected if one charges for it?". In this section some information and parameters to be considered in the development of health care financing schemes will be grouped into three categories: recurrent costs, demand for care, willingness and ability to pay. Others will be mentioned in later sections.

Recurrent costs. To estimate the recurrent costs of existing health services an inventory of all facilities and health programs currently in operation is necessary. Fixed operating costs and semifixed costs, mainly for personnel, must be estimated based on present budget allocations. Some idea about variable costs for medical supplies and drugs can be derived from current service utilization rates, even though these are unlikely to reflect the demand for care (see below) and unit prices.

The future expansion of health services and the development of unit costs has to be known so as to be able to project recurrent costs for some years. Such information is usually found in five year development plans.

The computer model developed for the World Bank (Barnum and So 1986), implemented and applied by Shepard, Kleinau and Barnum (1990), allows an aggregation of these data and a projection over 10 years. Current costs for services and the effect on demand can be calculated in detail with the model developed for the REACH project (Bitran and Deal 1988).

Demand for care. The demand for care can be derived from demographic, socioeconomic and disease incidence data and only approximated from current utilization rates for populations which have adequate access to quality care. Some studies have attempted to determine the demand for care in Zaire and Nigeria. Definition of demand is a normative process, which very much depends on localized experience, on one's professional opinion, and on what is considered appropriate for a given disease incidence.

Factors influencing demand are *prices of care, degree of price discrimination, risk sharing, quality of care, competition, accessibility,* and *income.* Price elasticity of demand has been studied in the papers by Shepard et al. (1987), Carrin (1986) and Vogel (1989).

Willingness and ability to pay. This capacity of the target population is usually determined during demand for care studies. The political willingness to allow alternative health care financing schemes beside the government and donor funding can only be assessed in a country.

Current utilization rates of government health services are poor estimates of demand for care. As mentioned, these services are frequently underfinanced and deliver a bare minimum of services. The quality of these services is often inadequate: supplies are insufficient, structures are poorly maintained, personnel is disgruntled. Even though private facilities might require greater expenses for travelling and services they are usually much more used because quality of services is perceived as high. Their utilization rates and costs might provide better estimates.

In the centralized public sector health institutions do not act as rational providers of care by adjusting their inputs to demand. Outputs, (services delivered), will produce misleading results when related to resources employed through a production function. The private not-for-profit and more the for-profit sector can be expected to act as rational buyers of inputs in a free market.

An Expert System Modelling
the Selection Process of Health Care Financing Schemes

General Considerations for Design and Implementation

If the process of selection of health care financing schemes is so much dependent on the political environment and has many uncertainties built in, why bother to develop an ES? The ES cannot rationalize the political process. One can, however make the assumption that alternative policy options are ignored because they are not presented as such. An ES could increase the awareness for a cost-based rather than an expenditure-based approach, show different schemes of cost recovery, and demonstrate their potential contribution to health care financing.

An ES can reduce the complexity involved by incorporating basic data, the relationship and interaction between parameters. This would ease the evaluation of health care financing schemes for busy (and with respect to economics and health care delivery) less-experienced decision-makers. One substantial advantage of an ES could be that it does not need to be biased towards certain solutions. Rather it can present the full spectrum of schemes guiding the user to pick the most appropriate ones under a given scenario.

The position of an ES in the Anthony & Simon matrix has implications for the design and implementation of the ES. Keen (Keen and Morton 1978) identifies four areas which are influenced by the characteristics of the problem structure and the management level for which an ES is designed: people, models, process, technology.

People involved in building an ES for semi-structured decisions ideally should have good technical skills and know the decision-making process very well. The latter seems even more important than the former when it comes to model building. The technical expertise to write a computer program for the ES could be brought in through a team of health economists, international development experts and computer programmers. System analysts without understanding of health care financing schemes and the decision-making process will probably not effectively design a specific ES.

Building a model for the decision-making process requires a different approach than for structured accounting or manufacturing problems. Sophisticated algorithms for optimization, like linear programming, are not appropriate in a situation with many parameters based on scanty information and poorly understood interactions. Keen points out that small, informal models provide better answers. A simulation model reflecting the approach of decision-makers in ministries or donor agencies, but enabling them to make better decisions is preferable to abstract econometric models.

The process of developing an ES is evolutionary. If the ES is to have any chance of being used then the active involvement of the end user in the design and development is crucial. The focus on two different user groups with different objectives for their decision-making might pose a problem. Due to the need for an open-ended approach the ES should be easy to expand and be possibly modular. Different modules would relate to recurrent costs, effects of various financing schemes, country specific information and political factors.

The technology either of stand-alone computer or a network of micro computers is suitable for running an ES. Users are few as compared to data processing for operations control and an ES is used infrequently, except for the entry of baseline data. This type of low cost computing is available today in the Ministries of Health of most developing countries or they have access to them through donor agency projects. As an output device a LCD display for overhead projectors together with appealing graphics will make an impressive presentation of ES recommendations to high level management and politicians.

People: User Reaction and Relevance of the Solution

Some of the management literature suggests that senior managers do not use computers extensively. Rather they spend their time interacting with people directly (Rockart 1979; Kotter 1982). However, management scientists do not share this opinion unanimously especially since a decade has passed since many such statements were made. This problem might be circumvented by designing an ES with a user friendly interface and easy data entry which can be performed by lower-level personnel. This would make it more a presentation, planning and training tool, which might augment its acceptability. It is assumed that microcomputer technology is available and training capacity sufficient, at least on central levels in the health sector, in most developing countries.

The reaction of politicians to more information is hard to judge. Some might find it an unnecessary nuisance to be presented with unwanted choices, especially as administrative and organizational consequences can be required to make the alternatives work, such as decentralization and privatization. Others might welcome means to improve health care delivery. The health service faces a similar ambiguous situation. There is a number of committed and even entrepreneurial professionals who would make good use of improved financing by providing more and better services. The less motivated civil servant might resent additional work. These considerations make strong leadership even more important – leadership to use a computer based ES to render decision-making more effective, and leadership to apply the recommendations.

This paper will not deal with aspects of behavioral research: people, personal needs, the social system and work environment, resistance to ES and computer technology in general. So far the ES does not seem to affect many people in the organizations considered here, assuming that computer technology is accepted already.

Anticipating that the solution offered by the ES is relevant does not answer the question whether the benefit of more information is worth the costs (to develop the ES and user time to enter data, retrieve and interpret information). It is probably safe to say that the ES produces no cost savings in the form of less personnel or time needed for making decisions. Instead even more time by lower level staff might be required for data collection.

Model Building

Several aspects should be considered when modelling a decision-making process. As mentioned already, a semi-structured problem embedded in a complex management environment seldom allows the application of an existing model. Familiarity with the situation is probably as or more important than technical know-how. The model has to be developed through observational studies; it should fit the actual process. At some point during the development of the ES a decision between a normative or a descriptive approach has to be made. In the case of selecting financing schemes, available information is too inaccurate to favour a normative approach. Even though some elements such as personnel per health unit or an expected utilization rate are normative, the overall approach is descriptive, reflecting limited experience and actual observed relationships instead of hard and fast rules.

For the above reasons, decisions for health care financing schemes are more heuristic (rule of the thumb) rather than based on optimization algorithms. This is not to say that such techniques are not useful for parts of this ES, when it comes to finding the "best possible" mix of financing schemes based on revenues and a number of constraints. However, the conclusions emerging from strictly quantitative analysis should be validated against past experience and theory.

The effect of several parameters, political acceptability of concepts, price elasticity, etc. have a high uncertainty and the decision for financing schemes is likely to be substantially influenced by them. To test the sensitivity of outcomes to such parameters and to predict future developments of costs and revenues, the ES contains a simulation component allowing "what-if" type analyses. Figure 13.2 shows an outline of the model structure as derived from previous discussions. Its elements are discussed in more detail below.

Intelligence. Intelligence is to a limited extent the duty of senior management. It is their role to formulate the right questions as related to health care financing and to make sure that the right data are collected and additional research is carried out. The actual data collection and the updating of the knowledge base and the parameter bank should be done by junior staff.

Contents of the knowledge base. As much information as possible accompanies the ES, requiring the user only to update, correct or add new data. With less knowledge incorporated, the ES would rely more on guesses, rendering its recommendations less reliable. The possibility to update the data base annually as part of supporting and maintaining the ES should be evaluated. Data for the knowledge base are found in World Bank, WHO and country publications, most of which are publicly available.

The distinction between knowledge and parameters is somehow arbitrary. Knowledge in the context of health care financing, as defined here, refers to general characteristics of a country which are condensed to specific

FIGURE 13.2 Financing Scheme Portfolio Model Structure

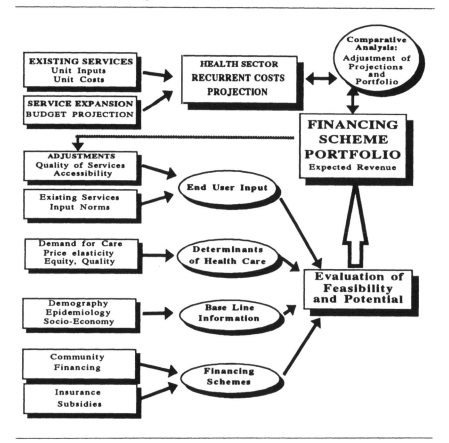

indicators influencing the decision about financing schemes. General economic developments like world market prices for important export products of developing countries, characteristics of financing schemes in general, or standards with wider applicability like proportions of medical personnel for different types of facilities would also fall into the knowledge area. Rules are frequently treated as part of the knowledge base in artificial intelligence literature, but for clarity they are discussed separately in this paper.

Parameters express a relationship, an effect, or they are ratios such as population based service utilization rates and fatality rates among hospitalized patients.

Parameter sources and input. Some parameters have wider applicability or the same sources are referenced because few exist (demand for care, price elasticity, socio-economic variability). These are built into the model, presuming medium term validity for the more static parameters. Other parameters show a more dynamic characteristic (relative prices for services and supplies, political willingness). Accordingly, these are provided by the user.

A number of parameters are easily accessible from service records like hospital care utilization, fatality rates among reported inpatient cases, or costs per patient, especially since some countries have improved their data collection and evaluation systems by introducing microcomputer technology (Rwanda and Togo among others). Some parameters will be found only in special studies (demand for care, price elasticity, regional household income, disease incidence), and some will rely on intuition and experience (political factors, economic development).

Data aggregation and subset creation. Data from the knowledge base need to be further aggregated to be useful in the evaluation of health care financing schemes. Based on population size, disease incidence and average hospitalization per disease episode the expected service utilization rate can be calculated. It can later be adjusted according to health infrastructure, quality and price of services. From the average household income and its distribution among population groups the potential to pay for health care can be derived. The model should be applicable to any geographical subset, for example, rural versus urban.

Knowledge Base. In artificial intelligence applications like an ES, knowledge – including parameters – can be represented in several ways, as described by Harmon and King (1985). Most useful for a health care financing application seem to be the following.

- *Semantic networks* represent baseline information from different developing countries. A country is assigned to a node, as a conceptual entity or object, which is linked by an arc to other objects like per capita GNP or health expenditure. The link can be translated to: "A country *has-an* annual per capita GNP", "The annual per capita GNP *is* 200 US$"; or "A country *is-a* SubSaharan nation".

- *Object-Attribute-Value (O-A-V) triplets* can be considered as a special case of a semantic network where a *has-a* link is followed by a *is-a* link similar to the first example above. The country and the existence of an annual per capita GNP is static information, the actual value is dynamic.

- *Frames* present knowledge similar to the O-A-V triplet, but in a richer form. A complete list of attributes and values is attached to a single object, some of which carry default values.

Rules and Decision Tree. Rules are needed to combine knowledge and parameters to recommend certain financing schemes and to highlight advantages or disadvantages. Due to the inherent uncertainty of much of the information, a decision tree maximizing the probability of an outcome is used to make the best recommendation possible under these circumstances. Rules can process any type of data as long as their meaning is not ambiguous. For most operations it will be necessary that parameters are numerical. If they are qualitative, such as "Support of user fee" or "Willingness to pay 10 US$ insurance fee", a simple scale can be used and converted into numerical values, assuming that more precise information is not available, for example:

Rating scale:
1 = none 2 = little 3 = about 50% 4 = substantial 5 = full

Such a response curve can be further refined to any appropriate level and is preferable to the direct index approach. If parameters have to be estimated, it would be easier to ask for values relative to a "standard" setting and arrive at ratings like better or worse than this, instead of asking for absolute classifications. A value would have to be assigned to the standard, 1 or 100 for example. Such a procedure would be suitable to evaluate service intensity or the quality or care.

If conclusions are not certain, one assigns a certainty factor between 0 and 1 to such an outcome, zero indicating an impossible outcome and one a certain one (Winston 1984). The way overall probabilities for several rules in series or parallel are calculated is less obvious. One can apply probability rules of addition and multiplication or deviate from them based on perceptions about the reasoning behind the decision process.

For example, if one is 60% certain that the government will approve financial autonomy for hospitals and there is an 50% chance that this will lead to quality improvements resulting in a 30% increase in demand for ambulatory care, then it will be of less interest how much on average we can expect service utilization to increase. Will the probability be 30% to see a 30% increase or will there be a 60% chance to see a 15% increase? Both will result in an average of 9%, but implications for decision-makers would be very different. In the first case one would hesitate to make provisions for higher volume, because of a low probability. In the second case one can decide the opposite, but envisage a more modest increase of volume.

Rules are further refined by propositional logic, where a statement such as "Private saving clubs exist in a society" are either TRUE or FALSE. Several statements can be connected by logical operators like AND and OR. The conclusion will depend on the overall truthfulness of such compound statements.

A complex set of rules can be combined in a decision tree with decision and chance nodes if monetary benefits, penalties or other forms of utility are assigned to outcomes. In keeping with the former example, if volume actually increases and if key personnel decided to increase supplies a relatively higher proportion of total service costs might be covered than under the *status quo*. If, however, supplies were increased and neither political change nor better service quality did occur then excess supply could further deteriorate the financing situation due to losses and expensive storage.

Drawing Inferences. The inference engine, in ES terminology, analyzes information contained in the knowledge base, the parameter bank and the rules. It then proposes a solution for the underlying problem, finding an optimal financing scheme portfolio. The ES arrives at its recommendation through different methods of reasoning. The user can influence the outcome by providing additional information. The order in which inferences are drawn depends on intermediate results during the process of evaluating facts and rules.

This ES like most Expert Systems makes inferences following the *modus ponens* following an "IF...THEN" logic as outlined in Silverman (1987). He distinguishes at least two levels of rules: those at the top deriving the final decision, and those on a second level providing specific outcomes necessary for decisions on the higher level. The second level consists basically of four major rules: the relevancy, feasibility, optimality, and success rule. Simplified examples applying to health care financing will follow.

Relevancy rule: IF curative services lack cash to finance the provision of essential drugs THEN selling drugs to patients is a relevant solution.

Feasibility rule: IF evidence from the private sector shows patients are paying for drugs when required AND IF average family income appears high enough to support a greater share of expenditures for health care THEN selling drugs is a feasible solution.

Optimality rule: IF drug sales cover a high proportion of costs AND IF it is easy to implement AND IF it can be equitable (through price discrimination, for example) compared to alternative health care financing schemes THEN drug sales is the best possible solution as compared to alternatives. The optimality rule is used in a more moderate way than described in the literature. The word "optimality" seems to be driving more at the "absolute best solution", which would be counterintuitive due to the uncertainty inherent in the problem of health care financing. In the context of health care financing it might better be called the *best possible solution rule.*

Success rule: IF managerial capacity is developed AND IF revenue generation is adopted as government health policy AND IF government institutions are granted sufficient financial autonomy THEN drug sale is a successful solution.

As mentioned in earlier sections, some outcomes might not be absolutely certain, but they can be expected with a certain probability. This could influence the outcome of the *top level rule* which reads: IF drug sale is a relevant AND feasible AND optimal AND successful solution THEN drug sale is the appropriate solution. If the overall appropriateness of drug sales has a certainty factor below one, other solutions can achieve a higher score and be more appropriate. This two level rule set is expanded to more levels in the actual ES application. The principle applies to any number of levels.

The inference engine has to make an important decision: where shall the process of reasoning start? The order is controlled according to a forward or backward chaining process and with a *depth-first* or *breadth-first* perspective. *Backward chaining* through the rules is appropriate if not many financing schemes are to be considered by the ES. In this case each final solution would be tested for its consistency with lower level rules and facts. If many schemes have to be considered and many conflicting situations have to be resolved the *forward chaining* process is more appropriate, where one departs from basic facts and works towards solutions.

In a *depth-first* approach the inference engine follows a single financing scheme through all levels of the rule set deriving a single recommendation, ignoring alternatives at first. This can create a problem if, in an attempt to propose a solution, it becomes impossible for a number of alternatives to pass one or several critical IF...THEN rules. In this case a *breadth-first* approach could be more efficient, because all possible alternative financing schemes are tested on one level before proceeding to another level. The chance of passing any rule successfully is greater.

Choice. Ranges and sensitivity analysis. Because of the inherent uncertainty the model cannot provide absolute and precise figures. Reasonable ranges should be available for all indicators, utilization, demand, price elasticity and outcomes, or revenues generated per scheme. This provides the user with some idea about the variability, but still conveys more information than gut feeling.

This approach should be extended to an automatic sensitivity analysis, which tells the user at which values the outcome would change and how far these are from the 95% confidence interval. This would be similar to identifying slack resources and dual prices in linear programming.

Presentation of alternatives. The user has a choice between a complete list of all available financing schemes under one scenario, a short list of possible or recommendable schemes and a recommendation based on revenue maximization. Each scheme comes with information related to revenue, feasibility, acceptability and effects on equity, quality and other factors.

Development Process of the ES

Several concepts exist for the description of the decision-making process, each influences the design of an ES differently (Keen 1978):

- The economic rational concept (looks for the best solution).
- The satisficing process-oriented view (find a good enough answer).
- The organizational procedures view (focuses on interrelations within the organization and standard operating procedures).
- The political view (game theory, process of bargaining and conflict).
- The individual differences approach (personal style of decision-making, personal values, subjective).

Considering the discussion up to this point, the satisficing process-oriented view seems to suit a situation with considerable uncertainty. The political view, even though important cannot be built well into the ES. Figure 13.3 presents an overview over the structure for the Financing Scheme Portfolio ES. As in previous sections, the framework of intelligence, design and choice as set forth by Bennett (1983) will be used for the analysis of the decision-making process.

FIGURE 13.3 Financing Scheme Portfolio Expert System

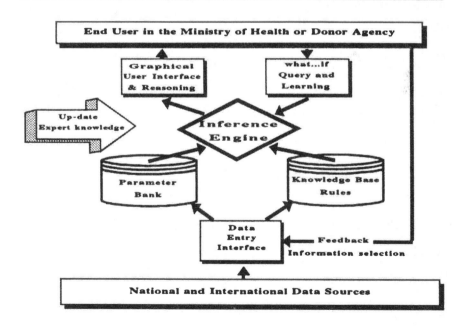

Implementation of the model, testing, refinement. Close collaboration with end users has been emphasized. From an early stage the development of the ES should be associated with a donor agency and possibly with an interested Ministry of Health through one of its projects. This user involvement would hopefully create the necessary curiosity and eagerness to try out the ES and evaluate its performance.

Performance and completion criteria. It was stated earlier that an ES is never completed. Nonetheless, it is important to define criteria as to when certain *phases* of the development can be considered completed. A first prototype should be developed rapidly, within several months. It should contain at least 80% of the planned features (assuming that the 80/20 rule applies here, it would mean that 80% can be achieved with 20% of the effort and input required). The error rate concerning program functioning probably should not be greater than 10% so as not to frustrate users, who would cease using the ES.

For the Financing Scheme Portfolio ES one end-point definition could be the development of the knowledge base, the parameter bank, the rule system and the graphic user interface. This prototype would make already meaningful recommendations. The validity can be tested by analyzing historical data with known outcome. More advanced features such as the English language report, explaining the decision-making process and learning from the user could be implemented in later phases.

Continuous development, calibration, maintenance and support. Close communication with users has to be maintained to correct errors in a timely manner and to enhance the ES. Because health care financing is a complex topic in itself the ES software (computer disk and manual) is probably not suitable for simple distribution without user training. Potential users should be introduced during seminars concerning health care financing for which the ES is used as a learning tool. The "chauffeur" approach, where a consultant operates the computer and executes the user input, should not be necessary, because the design of the ES is aimed at user friendliness.

The model can be further calibrated by changing the interaction and influence of parameters during a trial period. The following means of calibration could be used: judgement of appropriateness, analysis of historical data, tracking (historical data analysis with feedback loop to adjust the model), field measurement and adaptive control (observe actual implementation of health care financing schemes and use ES for parallel evaluation, adjusting the model according to observations).

Control and evaluation. The effectiveness of the decision-making process prior to the introduction of the ES should be formally evaluated and documented. This would allow a comparison with an evaluation carried out some time after implementation. Criteria for what is considered effective decision-making and a successful outcome should be also defined.

Design and Technology

Keen's recommendations concerning basic design principles of a decision support system also apply to building an ES:

- An ES should be simple and user oriented. It should include relevant information but rely on approximations rather than precise academic models. Aggregate response is preferable to micro-simulation, which might be wrong anyway.
- It must be fast (response time), old PCs and mainframe computers might be unacceptable.
- The ES must be foolproof.
- It should reflect the management process.
- User should be involved in planning and design.
- Output has to be graphic oriented.
- The ES should aggregate information which, in its raw format, would not be available to the user.
- The ES should encourage learning and foster alternative judgements. It should allow subjectivity on the users side.
- The ES should be constructed in modular form, more detail and specification should be dealt with in sub-modules.

Graphic oriented user interface, menus, context sensitive help. The end user should be required to type as little as possible, point and click or touch devices are preferable. A simple and clear screen layout can make the ES appealing or repugnant if cluttered. A LOTUS 123 layout with a menu displayed in the top rows with a brief explanation of menu options and a 20 row display section is one appropriate solution. With the Vector Graphics Array (VGA) becoming the new display standard more information can be presented on screen and at a higher quality.

Frequently seen are Graphical User Interfaces (GUI) as used by MS-Windows or on the Apple Macintosh type pull down menus. Care must be taken, however, that the user does not have to go through too many layers of menus to enter data or perform operations. Shortcuts must be available in each case. A context sensitive help provides information on program operation, explains how the model is functioning, and shows the flow of information.

Interactive graphs and tables, used with pointing devices, would greatly enhance the presentation. They allow certain information to be displayed instantly, changing between tables and graphs without going through menus. A printed English language report summarizing the basic scenario, parameters and recommendations would render the life of end users easier as well. It should include relevant graphs and tables.

For presentation purposes animated pictures, in form of a slide show or film like, are very attractive. Technically this has become feasible, even though requiring substantial computing power beyond the first generation of personal computers. In a good presentation the audience should identify itself with the product (the ES). A presentation of facts alone will not achieve this. Introductory screens, explanations, menus, graphs and pictures should reflect the political and professional culture of the country or the agency to which the ES is applied. Insignia, leitmotifs or wisdom of political leaders are such elements (remember, computers do not yet have emotions). This interface has to be very flexible and easy to adapt, which makes it a challenge to programmers. The day to day user of the ES should be able to turn this feature off to be able to concentrate on substantive issues. For presentations, however, such a feature could make all the difference between keeping the audience at the edge of their chairs or boring them to death.

A true expert system has two additional features: it is able to explain its reasoning and is able to develop its decision-making power by accepting new information and rules from the user. Such features need not only much more elaborate programming, but they also require more expertise from the user. The user communicates with the ES in near English language instructions aided by lists of options rather demanding him or her to memorize commands.

The ES should suggest default procedures for the "what-if" and sensitivity analysis to the user, which represent those most frequently used. These can be overwritten if a situation requires special analyses.

AI, third generation language, or standard application software. Several software alternatives should be considered. LOTUS 123 would be appealing because of its widespread use and its elaborate macro language. The graphic output is acceptable. The space is, however, very limited and the model is too complex for LOTUS 123 to deal with it in a user friendly manner. More appropriate alternatives are third generation programming languages such as "C" or Pascal. Turbo Pascal from Borland has the advantage of common use.

Extensive graphic, mapping, spreadsheet and database management libraries have been built for these languages which can be easily linked with other source code. This reduces the otherwise extensive programming effort. Parts of the ES related to problem definition, searching and linking different pieces of knowledge can be developed more efficiently with an artificial intelligence language such as Prolog or LISP. These are combined with conventional programming of algorithms. However, more research is necessary to identify application software or general purpose expert systems allowing a fast development of the ES. A new generation of object oriented programming languages for personal computers promises to expedite ES development.

Technology: hardware considerations. Personal computers are available in most developing countries and would be suitable to run the ES. More important are processor speed and input devices, especially if complex graphics and animated pictures are used. A mouse, trackball, digitizer (if maps are included to highlight regional differences), or a lightpen should be present apart from the keyboard. Convenient output devices are matrix or laser printer and a color plotter. Laser printers are becoming a serious option in many developing countries, assuming resources for operation and maintenance are available.

Description of Required Output

Knowing the general nature of the output before working on technical problems is a useful procedure to solve complex models. It also reduces the risk of getting lost in unwanted detail. According to knowledge base contents and parameters presented in previous sections some of the expected output from the Financing Scheme Portfolio ES is described in the following paragraphs and figures.

Recurrent costs (TRC) are shown per health care facility and program (Figure 13.4). A straight calculation of unit costs (UCF, UCP) times number of facilities (F) or programs (P) is adequate.

$$TRC = \sum_{i=1}^{n} UCF_i{}^*F_i + \sum_{j=1}^{n} UCP_j{}^*P_j$$

i = category of facilities, j = category of programs

FIGURE 13.4. Annual Recurrent Costs for All Facilities by Type

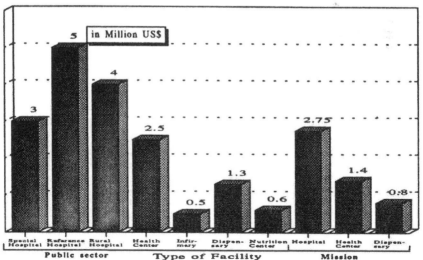

The unit costs are composed of fixed, semi-fixed and variable costs. Semi-fixed costs like personnel can be derived from a staffing standard per health unit and personnel category related costs, like wages and benefits.

Variable costs can be calculated from the expected *utilization* and the supplies, materials and drugs used per client. Drugs should be a separate item in the model for each facility and program to be able to calculate revenue from drug sales and remaining costs to be covered by other sources.

The costs calculated in this manner can only represent an average for a certain type of facility or program. Within each category such as rural hospitals variations can be considerable in actual size and the range of services offered. To capture this variation without burdening the ES user with detailed data entry demands, *ranges* for fixed, semi-fixed costs and variable costs can be specified. Such a range should cover costs as encountered in 95% of all facilities of the same type (mean ± 2 Standard Deviation). Three figures will be provided for total costs: a lower limit assuming all lowest unit costs, the average, and an upper limit assuming all highest unit costs. This is shown in Figure 13.5 for costs covered by user fees. The concept of ranges rather than precise estimates is useful for other parameters as well, such as utilization of services and demand. In case ranges turn out to be unacceptably large, one category might actually represent different subgroups for such facilities or programs. They would be analyzed separately. The decrease in costs recovered from fees is due to the implementation of new programs which do not generate revenue, but add to recurrent costs.

FIGURE 13.5 Percent of Recurrent Costs Covered by Fees for Services 1988–1997

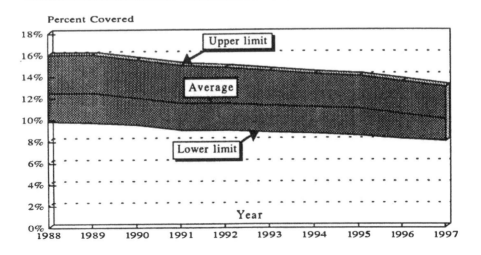

So far recurrent costs were compared to the *sum of all revenues.* To identify the contribution by source revenues must be broken down by facility and program category as were costs (Figure 13.6). A lower and upper limit and the average will cover most situations.

The difficulty lies in the allocation of revenues to each category of facilities or programs for each different financing scheme. In some countries the government budget will be very explicit, but may be on a regional basis rather than on unit basis. The amounts for the same category might vary, especially when political or equity considerations play a role. The user would have to figure out ranges and averages. A scratch pad area of a spreadsheet with the appropriate statistical formulas entered is helpful to avoid hand calculations.

With other health care financing schemes an allocation might be easy. Revenue from drug sales or user fees go directly to the health unit. HMO type settings or insurance schemes with several levels of service provision allocate according to costs.

Some of the schemes will depend on demand, approximated by the expected *service utilization* (Figure13.7) and the *market share* (Figure 13.8) of each category of facility (this is probably less of a problem for programs).

FIGURE 13.6 Financing Portfolio Recurrent Costs Covered by Source

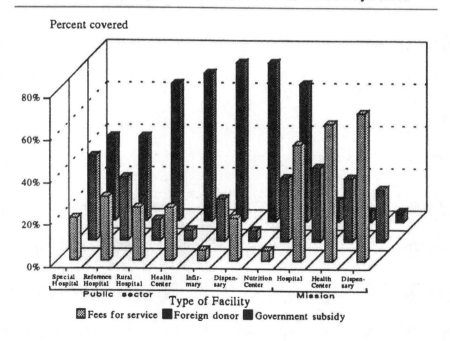

FIGURE 13.7 Monthly Utilization of Services

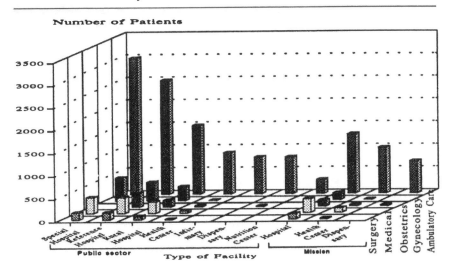

The proportion covered by *competitors* should be known to estimate later the effects of pricing and quality of care. The existence of a *referral system* will also influence utilization. Without such a system increases in prices can cause undesirable shifts among levels of care, from primary to secondary or even tertiary for example. It is a challenge for the ES to estimate these effects in an aggregate form rather than for individual units.

FIGURE 13.8 Market Share of Facilities by Type and Sector

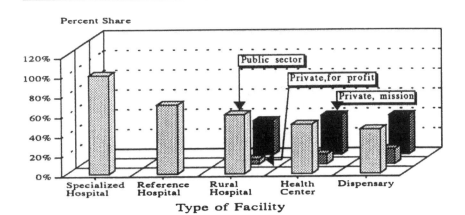

FIGURE 13.9   Costs per Disease Episode

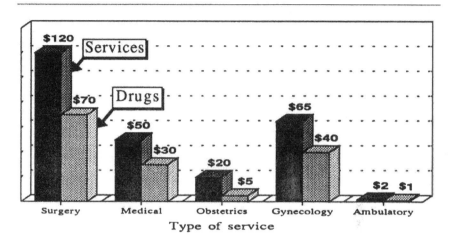

Another feature of the ES lies in its capability to suggest global *user fee revenues* to balance an eventual gap between costs and all other sources of finances for each facility and program. Such global fee can be broken down according to the costs per type of service rendered (Figure 13.9). Price discrimination can be implemented by weighing types of services according to resources consumed or their value to promote health. This way curative care or private consultation could finance preventive care like immunizations. The user fees calculated this way can then be compared to limiting or favorable factors, income, political acceptability, equity considerations, and so forth.

FIGURE 13.10 Public Sector Budget Gap

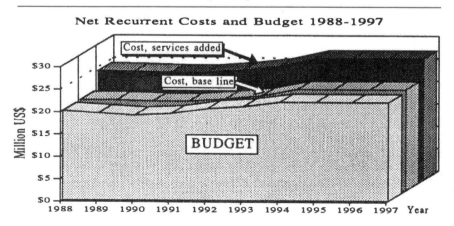

Net Recurrent Costs and Budget 1988-1997

For *health insurance* schemes the ES proposes premiums according to an expected number of subscribers, charges per person without insurance and the expected utilization rate, which usually differs from the uninsured population. Some reliable demographic information is necessary to do the calculations. Such schemes could include co-payments and deductibles to attenuate moral hazards such as excessive utilization of services. To be useful for planning purposes most data have to be projected for some future years, five to ten usually being sufficient. Figure 13.10 compares recurrent costs less user fees to the expected budget for 1988 through 1997. Figure 13.11 compares cost coverage by fees for each type of facility.

Improving Health Care Financing

Why bother with a computer model if decisions can be made without it? Certainly this question will be asked. Perhaps the time and funds should be allocated instead to more training of the personnel concerned. Training is certainly very important, but the ES is not an "either or" choice. Rather it would serve as an addition to enhance training, providing more information and more precisely than imparted by training alone. Also, it responds to time constraints of decision-makers and qualified trainers. Once the basic introduction is done the ES is available at any time, unlike trainers.

Developing the proposed financing scheme portfolio ES would be a very challenging task. One alternative approach would be to address this complex solution stepwise. This approach has its merits as it reduces costs and possibly avoids some of the pitfalls. Several areas have been or could be solved separately: assessment of recurrent cost for health services, a knowledge base incorporating experience and characteristics of health financing schemes, or modelling for individual schemes.

FIGURE 13.11 Annual Recurrent Costs and Revenues from Fees for Services

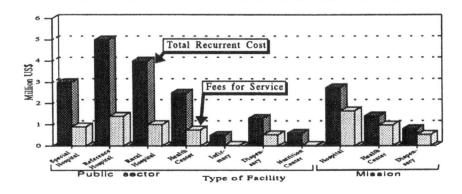

Conclusion and Critical Issues

Assumptions

One underlying assumption is that the decision-making process as described earlier has been identified correctly and that all relevant information has been obtained. Despite the fact that factors not based on hard data play a crucial role in health care financing, decision-makers are expected to have a desire for more and improved information provided through the ES. It is also presumed that they prefer to make better decisions. It is critical to verify these assumptions in direct interviews with key decision-makers. The Delphi technique would allow the development of qualitative aspects. New information has then to be incorporated in the model.

Potential resistance to implementation

The inculsion of political parameters could influence the acceptance of the ES negatively. First, political issues are always sensitive, especially to users who are acting at a much lower level in the political hierarchy and who can possibly be put in an embarrassing situation when trying to assess these factors. Politicians also may react negatively, because they might be confronted with results from the ES and they would rather not see a political decision materialize in printed form. Second, political factors are not easy to value or even to assign a number. For donor agencies this should be much less of a problem.

Social and cultural factors can be as sensitive as political ones. This might be less problematic if expert knowledge from sociologists and anthropologists familiar with the developing country were included in the ES's knowledge base.

Depending on the expertise available in the government, countries might resist such an ES, because knowhow in the form of human experts is available. The impression (wrong) that a computer model would actually prescribe decisions makes it important to circulate careful information about what the ES can do and what it cannot do. Those who operate the ES cannot be expected to have the power to influence decisions on financing strategies. Those who do have this power, however, might benefit from the additional information and recommendations provided by the ES.

In these cases the ES might still be acceptable as a planning tool for a team of experts, where these can feed their knowledge into the ES and compare their recommendations and reasoning to that of the ES. The ES is potentially very helpful in identifying, collecting and condensing important information that otherwise might be omitted. It also is likely that the introduction of the ES is more acceptable on a regional level, where it might be found less threatening and where the lack of human experts is more obvious.

References

Abt Associates, Inc. 1988. *A Household Health Care Demand Study in the Bokoro and Kisantu Zones of Zaire.* 1988. USAID, REACH Project, Cambridge, MA.

Barnum, H. and J. So. 1986. Health Finance Planning Model. The World Bank. PHN Technical Note 8629.

Bennett, J.L. 1983. *Building Decision Support Systems.* Reading, MA: Addison-Wesley Publishing Co.

Bitran, R. and D. Deal. 1988. Notes on a Financial Model of a Zairan Health Zone. Abt Associates, Inc.

Bossert, T. et al. 1987. *Sustainability of U.S. supported health programs in Honduras.* Washington, D.C.: USAID.

Carrin, G. 1986. The economics of drug financing in Subsaharan Africa: The community financing approach. Takemi Program in International Health, Harvard School of Public Health, Boston.

Gorry, G.A., and M.S.S. Morton. 1971. A framework for management information systems. *Sloan Management Review.* 13(1):55–70.

Harmon, P. and D. King. 1985. *Expert Systems: Artificial Intelligence in Business.* New York: John Wiley & Sons, Inc.

Keen, P.G.W., and M.S.S. Morton. 1978. Decision Support Systems, an Organizational Perspective. Reading, MA: Addison-Wesley Publishing.

Kotter, J.P. 1982. *The General Managers.* New York: The Free Press, MacMillan, Inc.

Rockart, J.F. 1979. Chief of Executives define their own data needs. *Harvard Business Review.* 57(2):81–93.

Shepard, D.S., E.F. Kleinau, H. Barnum. May 1990. Bridging the gap. Application of the health finance planning model in Rwanda. Presented at the Annual Meeting of the National Council for International Health (NCIH) Washington, D.C.

Shepard, D.S., and P. Nyandagazi. 1988. Topics in financing health care in Rwanda: unit costs, rural pharmacies, and the impact of AIDS. Rwanda: Ministry of Public Health and Social Affairs, and Washington: World Bank.

Shepard, D.S., G. Carrin, and P. Nyandagazi. 1987. Self-financing of health care at government health centers in Rwanda. Cambridge, MA: USAID, REACH Project. Harvard Institute for International Development.

Silverman, B.G. 1987. *Expert Systems for Business.* Reading, MA: Addison-Wesley Publishing Co.

Vogel, R. 1988. Cost recovery in the health sector. Selected country studies in West Africa. Washington, D.C.: World Bank.

Vogel, R. 1989. Recovery in the Health Sector in Subsaharan Africa. Tuscon: University of Arizona.

Winston, P.H. 1984. *Artificial Intelligence.* Reading, MA: Addison-Wesley Publishing Co.
World Bank. 1988. *World Development Report 1988.* Washington, D.C.: Oxford University Press.

14

Design of an Expert System for Water Pollution Determination/Prevention

Metka Vrtačnik, Pavel Čok, A. Cizerle, D. Dolničar,
S. Glažar, and Radojka Olbina

Introduction

In the last few decades, protection of surface and groundwaters from pollution has received a high priority since many countries have suffered from considerable deterioration in the quality of their water resources. This pollution is caused by discharges of untreated industrial and municipal wastewaters, agricultural run-offs, spills of hazardous substances, or "night dumpings" of industrial (hazardous) and municipal waste.

Data on river water quality parameters are gathered by laboratories worldwide. However an effective method of extracting value-added information from these data to facilitate decisions on the implementation of cost-effective pollution prevention measurements to safeguard people, livestock and industrial development in a region remains a problem.

While not always sufficiently recognized, artificial intelligence approaches based on induction learning and pattern recognition methods, represent a great potential in searching for regularities and patterns in large multiparametric data

The authors wish to express their appreciation to the Ministry of Research and Technology of Slovenia and the U.S. Environmental Protection Agency, Cincinnati, OH, USA, for their grants in partial support of this work. The authors thank Professor Aleksandra Kornhauser, for her professional support and help. The authors also wish to thank Slovenian Water Works Association, who allowed them to used data collected in their study.

sets. Commercially available AI systems (è.g. expert system shells) are designed to support searches for solutions in various classes of problems, e.g. classification, based on shallow, as well as deep knowledge, process design, process optimization, diagnosis, analysis, planning/scheduling, monitoring, process control (Gevarter 1987).

In searching for cost/effective solutions of water and wastewater treatment and pollution prevention, the following computer supported methods are emerging to extract value-added information from existing experimental data, often organized as global relational databases (Iwata 1989; Bengston and Parsmo 1990). For computer supported water quality modelling and pattern recognition approaches based on different classification models, see: James 1984; Fu 1985; Juricskay and Veress 1986; Wienke and Danzer, 1986; Van der Voet, Coenegracht, and Hemel 1987; Saaksjarvi and Khalighi, and Minkkinen 1989; Tiwari and Ali 1989; Gao and Xu 1989; Frank and Lantari 1989; Rump 1990. For induction systems based on inference theory of learning. see: Barnwell, Brown, and Marek 1986; Hushon 1990; Venkataramani, Bamopoulos, Forman and Bacher 1989; Perman 1990; Wable et al. 1990; Wood, Houck and Bell 1990; Seys, Duran, and Sivitz 1989; Ouzilleau, et al. 1990.

Problem Definition

The classification of Slovenian rivers according to their pollution level is defined by four categories (YU Official Gazettes No. 6, 1978):

Category I   (drinking water or drinking water after treatment)

Category II   (water that can be used for bathing, water sports or fishfarming, and could be used for drinking after adequate treatment)

Category III (water that could be used for irrigation or for industrial purposes after adequate treatment except for the food industry)

Category IV (polluted water)

The criteria for assigning a pollution category evel of a particular water sample at a sampling site are based on the results of: selected physical and chemical analyses; specific analyses of heavy metals and selected organic pollutants; and microbiological analyses. The overall pollution category of water samples is assigned by an expert panel which includes chemists, biologists and hydrologists.

Baseline data on water quality collected at 131 sampling sites for three consecutive years show that the quality of water is decreasing in the majority of Slovenian rivers. In most cases this is due to uncontrolled discharges of untreated industrial wastewaters and agricultural run-offs (Research of Slovenian surface water quality for the year 1986: The Slovenian Water Works Association, Hydrometeorological Institute of Slovenia, Biological Institute of Ljubljana University, Chemical Institute Boris Kidric, Institute for Public Health Maribor.)

To mitigate these problems, an attempt has been made to design an expert system to support a quick response to a water pollution emergency situation through:

- **identification** of selected pollutants with probable damaging effects, and in a given environmental situation with a hierarchical sequence of proposed tests to be carried out, taking into account the cost-benefit factor;

- **planning** immediate action to cope with an hazardous situation;

- **identification** of the actual pollution source(s), e.g. individual enterprises, farms, etc.;

- **providing advice** on ways to reduce pollution at the source in order to prevent a recurrence of the hazardous situation;

Such an expert system attempts to provide an efficient support of water pollution control at lower cost by providing sufficient reliability to define the water pollution levels based upon a reduced number of analyses needed. The system also attempts to provide advice for the most efficient wastewater treatment technology for a given stream.

Research Hypothesis

Knowledge in a given domain consists of descriptions that characterize objects, relationships, and procedures for manipulating these descriptions. In order to make expert systems useful tools to handle problems, a most appropriate knowledge representation must be searched for in a given problem domain, and an effective control of the reasoning process (which depends on the type of the domain and knowledge representation), must be implemented.

In order for a system to reason, it must be able to manipulate the data of a specific problem using the general facts of the domain (declarative knowledge) and the rules (procedural knowledge) until a satisfying solution is found or no more rules can be applied. (Klaessewns and Kateman 1987; Levine, Drang, and Edelson 1986).

One of the ways of representing knowledge of a domain is by means of relational or semantic diagrams. Thus, a relational database with general facts of the problem domain (declarative knowledge) was designed based on the methodological approach which is schematically presented in Figure 14.1. This relational database can be viewed as a template of a knowledge base of an expert system.

244

FIGURE 14.1 Outline of the Methodological Approach in Relational Database Design.

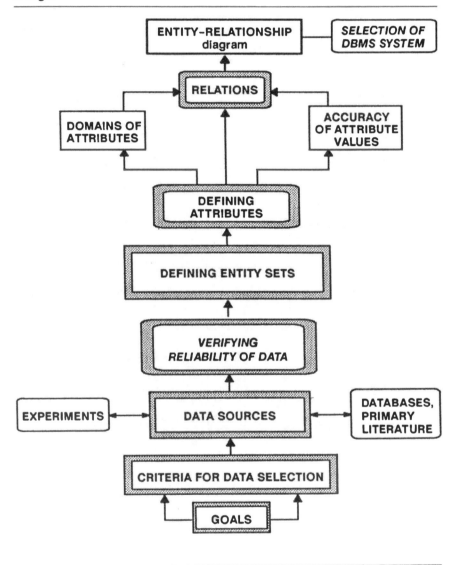

Design of Declarative Knowledge of an Expert System – As a Relational Database

Water pollutants included in the relational database were selected from the list of 299 compounds considered in Section 311 of the USA Water Pollution Control Act as compounds having a potential impact on the environment based on their toxicity and degradability in an aqueous environment. Approximately 250 compounds will eventually be included in the database. At present the database includes 128 compounds.

Sources of Information on Selected Pollutants

Several computerized as well as primary literature sources have been used in the collection of reliable information on: physical and chemical properties data, industrial occurrence and treatability data; data on biological effects (Appendix 1); analytical methods for pollutant determination in fresh and drinking water (Appendix 2); and site-specific data sources. For the Register of Water Bodies and Monitoring Points, sources are the Slovenian Water Works Association and the Hydrometeorological Institute of Slovenia. For Time Series Data on Chemical, Biological and Hydrological Aspects of Water Bodies, sources are the Slovenian Water Works Association, Hydrometeorological Institute of Slovenia, Biological Institute of Ljubljana University, Chemical Institute Boris Kidrič, Institute for Public Health, Maribor. For the Register of Slovenian Industry: Types and Location Coordinates, the source is the Statistical Bureau of Slovenia.

Defining Entity Sets and Their Attributes

The mathematical concept underlying the relational model is the set-theoretic relation, which is a subset of the Cartesian product of a list of domains. A domain is a set of values for a given attribute. A relation is any set of the Cartesian product of one or more domains. In order to develop the conceptual scheme of the relational data base model, first the entity-relationship diagrams are designed and then the real database systems are built on them.

The main elements of the entity-relationship diagrams are:

- *entity sets*   groups consisting of all "similar" objects (or actions, or concepts);
- *attributes*   (also called variables) used for describing properties of entity sets;
- *relationship*   a relationship among entity sets is an ordered list of entity sets.

In the process of the design of the relational database on water pollutants, the following entity sets have been identified:

- chemical substances (pollutants),
- synonyms,
- water quality standards,
- reactions in water,
- possible industrial sources of pollutants,
- other possible sources,
- wastewater treatment methods,
- cell multiplication inhibition test,
- test on acute toxicity,
- odor test,
- MERCK rapid tests,
- UV and visual spectroscopy,
- atomic absorption spectroscopy (AAS),
- thin-layer chromatography (TLC),
- high-performance liquid chromatography (HPLC),
- gas chromatography (GC).

Properties of each entity set are described with a selected set of attributes.

Entity-Relationship Diagram

The conceptual structure of the relational database model on "Water pollutants" is graphically summarized in the shown in diagram Figure 14.2. The model is designed in such a way that a series of general and specific questions can be answered by a proposed conceptual structure of the relational database:

- **What** are the probable sources of the pollutants?
- **What** are the observable biological effects of pollutants on aquatic organisms?
- **Which** are the main chemical/biochemical transformations of a particular pollutant in water?
- **How** can classes of pollutants be identified on the basis of the characteristic symptoms they cause, e.g., algal bloom, dead fish, odor, and so forth?
- **How** can a hypothesis on the possible cause of pollution be confirmed with the choice of a cost-effective analytical procedure for qualitative and quantitative determination of pollutants in river water?
- **How** can pollution be reduced at its source?

Design of a Site Specific Database

The conceptual scheme of the relational database on potential pollution sources in Slovenia, and time series of data on water bodies monitoring is designed in such a way that the following main questions can be answered:

- What are the levels of pollution of a particular Slovenian river body?
- How does the pollution level of a particular river body change over time?

FIGURE 14.2 Relationships Among Views, Tables, and Attributes

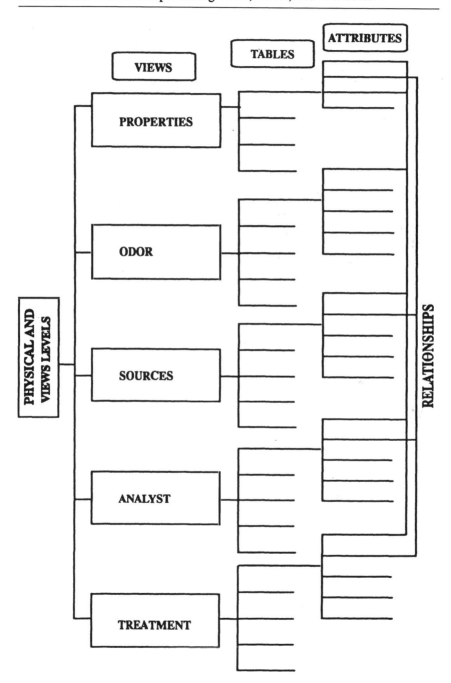

- Which enterprises/companies can be identified as possible sources of pollution at a particular monitoring point?
- How are pollution levels of Slovenian rivers distributed in regions/ communities?
- Where has the impact of river water pollution in Slovenian regions already caused severe damage to the environment and to industrial development?

The entity-relationship for Slovenian rivers and industries is presented in Figure 14.3.

Implementation of a Relational Database Model

Commercial database management systems that will fully support the entity-relationship approach are still under development. Thus, for the implementation of the entity-relationship schemes for "Water Pollutants" as well as "Slovenian Rivers and Industries" database models, the relational database management system INFORMIX-SQL, version 2.10.06C was chosen. INFORMIX–SQL is implemented on the industry standard definition for SQL and the proposed ANSI standard for relational database systems. INFORMIX-SQL system consists of usual programs or modules that perform data-management tasks.

The entity-relationships diagram for the relational database "Water Pollutants" model was transformed with the application of INFORMIX-SQL system into 18 tables. Each table corresponds to one entity set. The name of each column in the table represents one attribute. Tables were then joined logically into five multiple-table screen forms for data entry and data query by example:

- *properties*   properties of pollutants, potential industrial sources, biological effects on aquatic organisms and possible fate of pollutants in water,
- *odor*   odor quality and hedonic tone,
- *sources*   potential natural and man-made sources of pollutants in water,
- *analyst*   analytical methods for pollutants determination in water,
- *treatment*   wastewater treatment methods.

Design of an Expert System Knowledge Base

On the expert system level, facts in the relational database represent declarative knowledge, or a *global database,* while relationships embedded in the model enable recognition of elements of the procedural part of the knowledge base. In interdisciplinary multiparametric problems, relations among parameters

FIGURE 14.3 Entity-Relationship Diagram – "Slovenian Rivers and Industries"

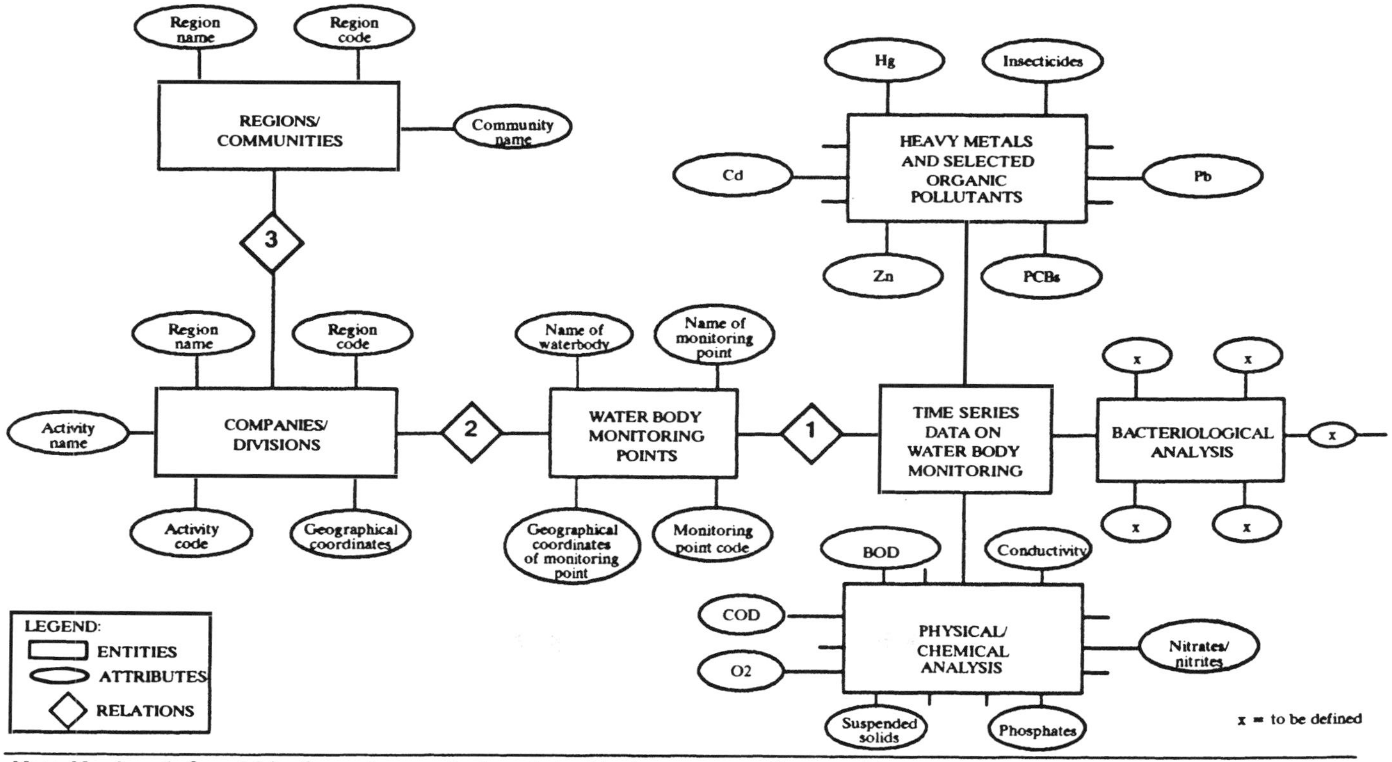

Note: Numbers 1, 2, and 3 in diamonds are codes for relationships between two adjacent entities.

cannot always be explained by the deep knowledge of the domain. Thus, processing data on the relational database and graphical display of relationships support the development of empirical rules. The level of generalization of the rule depends on the number and representability of examples in the database.

Generalization of the rules can be accomplished to a certain extent by a regular and reliable data input. The numerical data compilation is a limiting factor in quick developing relational databases. To overcome this problem, a subscription was made to the TDS-Numerica database. TDS-Numerica gives immediate access to up-to-date, reliable and evaluated data on properties of chemical substances with literature references. The most outstanding feature of Numerica is that properties can be calculated, based on other known properties or molecular structure, in cases where experimental data are not available.

The model of the expert system knowledge base is given in Figure 14.4. The line of reasoning of an expert system is not yet completely fixed, but at the present developmental stage it enables the user to check two hypotheses:

Hypothesis 1: The pollution is caused by discharges of untreated wastewaters from industries located along the river sites;

Hypothesis 2: The pollution is caused by spills of hazardous substances, which entered in the river by an accident or intentionally.

The goal of the consultancy is to identify, within a given region of river water pollution, the most probable industry or industries as pollution sources and to give proposals on the implementation of cost-effective wastewater treatment technology in order to minimize the river water pollution load. The logical relationships are implemented on the Knowledge.Pro communication system.

An integral part of the expert system is the classification rule, which enables the user of the system to compare the baseline data of the given river pollution levels with the actual results of analysis based on reduced number of measurements. This part is still under development.

Value-added information from the existing baseline data on monitoring Slovenian river water bodies was extracted to reduce the number of measurements for determining water pollution levels. For achieving this goal, discriminate analyses was tested in order to develop a reliable classification rule. The number of parameters which are used normally for the assignment of the pollution class could be reduced from 39 to as few as 21 predictor variables, with-

FIGURE 14.4 Logical Relationships of the Knowledge Base "Expert System for Water Pollution Determination/Prevention"

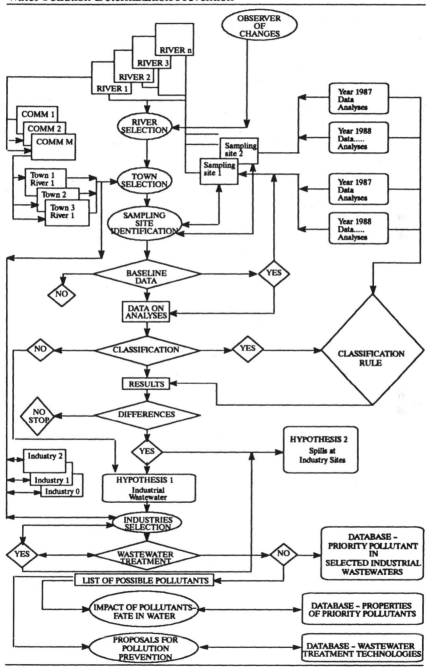

out significant decrease in the reliability of the classification score. (Vrtačnik, et al. 1991). Results were compared with the classification rule derived from the inductive expert system shell Assistant.Pro. Assistant is a member of the family of learning systems also called TDIDT (Top Down Induction of Decision Trees). This knowledge representation is in the form of a decision tree. The basic algorithm was developed from Quinlan's ID3, but improved to handle incomplete and noisy data, continuous and multivalued attributes, and incomplete domains. (Cestnik, Kononenko, and Bratko 1987). The classification tree developed by Assistant.Pro enables a determination of Slovenian river pollution levels derived from baseline data for physical-chemical microbial analyses. Due to the complexity of the classification tree, further work is oriented towards tuning the major criteria for the tree development to simplify its structure.

The classification rule will be evaluated, and if necessary, customized within a group of experts. The customized rule will be implemented on the expert system shell Knowledge.Pro.

Conclusions

This chapter describes a prototype expert system for river water pollution determination and prevention. The declarative part of the knowledge is represented in the form of two relational databases – "Water Pollutants" and "Slovenian Rivers and Industries." The reasoning mechanism of the expert system user interface is designed to analyze two hypotheses of possible causes of river water pollution (untreated industrial wastewater discharges and spills of hazardous substances). Further work is oriented towards the updating of existing relational databases and extraction of value-added information from the existing data by the application of commercially available induction systems, such as Assistant.Pro and Knowledge Maker.

References

Barnwell, T.O., Jr., L.C. Brown, V. Marek. 1986. Development of a prototype expert advisor for the enhanced stream water quality model QUAL2E. Athens, GA: Environmental Research Laboratory, Office of Research and Development, U.S. Environmental Protection Agency. Internal Report 87.

Bengston, U., D. Parsmo. 1990. Materials data handling. Interfaces between an expert knowledge system and user oriented problem solving. *Proc. Int. CODATA Conf., 11th Sci. Tech. Data New Era.* 102–106.

Cestnik, B., I. Kononenko, I. Bratko. 1987. ASSISTANT 86: A knowledge-elicitation tool for sophisticated users. In *Progress in Machine Learning.* ed. I. Bratko and N. Lavrač. Wilmslow: SIGMA Press. 31–45.

Fu, Guowei. 1985. Optimum estimation of river water quality model parameters using computerized scanning calculation graphic analysis gradient searching method. *Huanjing Kexue Xuebao.* 5(4): 373–85.

Frank, I.E., S. Lanteria. 1989. Classification models: Discriminant Analysis, Simca, Cart. *Chemometrics and Intelligent Laboratory Systems.* 5: 247–256.

Gao, H., Z. Xu. 1989. "Pattern recognition, method for water quality evaluation for the Songhua River," *Huanjing Kexue.* 10(1):75–7.

Gevarter, W.B. 1987. The nature and evaluation of commercial expert systems building tools. *Computer.* 24–41.

Hushon, J.M., ed. 1990. *Expert Systems for Environmental Applications.* Washington, D.C.: American Chemical Society.

Iwata, S. 1989. Expert systems interfaces for materials data bases. ASTM Spec. Tech. Publ. 1017 Comput. Networking Mater. Data Bases. 175–184.

James, A., ed. 1984. *An Introduction to Water Quality Modelling.* New York: John Wiley & Sons, Inc.

Juricskay, I. and E.G. Versss. 1986. Prima: A supervised classification method and its application in analytical chemistry. *Magy. Kem. Foly.* 82(6): 255–61.

Klaessens, J. and G. Kateman. 1987. Problem solving by expert systems in analytical chemistry. *Fresenius Z. Anal. Chem.* 326 (203).

Levine, R.I., D.E. Drang and B. Edelson. 1986. *A comprehensive guide to AI and Expert Systems.* New York: McGraw-Hill Book Company.

Miller, M.J. 1989. Verification and validation of decision support expert systems. Chemical process risk management in international operations. *ACS Symp. Ser. 408, Expert Syst. Appl. Chem.* 126–46.

Ouzilleau, F., J.B. Serodes, A. Beriault, and P. Suel. 1990. Application of expert systems to the management of accidental spills. *Sci. Tech. Eau.* 23(2): 153–9.

Perman, C.D. 1990. Improving the performance of wastewater treatment plants: an expert systems approach. *Diss. Abstr. Int. B.* 51(1): 407.

Rump, H.H. 1990. Matrix values in place of individual-substance values: possibility of evaluation of problem substances in wastewater. *Gewaesserschutz, Wasser, Abwasser.* 112 Abwasser-Abfalltech: Konzepte Prax. 331–49.

Saaksjarvi, E., M. Khalighi, and P. Minkkinen. 1989. Wastewater pollution modelling in the southern area of Lake Saimaa, Finland, by SIMCA pattern recognition method. *Chemom. Intell. Lab. Syst.* 7(1–2): 171–80.

Seys, S., M. Duran and M. Sivitz. 1989. Inteleau: Expert system for assistance in the treatment of sanitary hot water. *Eau, Ind., Nulsances.* 125: 35–7.

Tiwari, T.N. and M. Ali. 1989. Ground water of Nuzvid Town – regression and cluster analysis of water quality parameters. *Indian J. .Environ. Prot.* 9(1): 13–18.

The CRI Directory of Expert Systems. 1986. *HYDRO*, Oxford & New Jersey: Learned Information (Europe).

Van der Voet, H., P.M.J. Coenegracht, and J.B. Hemel. 1987. New probabalistic versions of the Simca and Classy classification methods. *Anal. Chim. Acta.* 192(1): 63–75.

Venkataramani, E.S., G. Bamopoulos, A.L. Forman, S. Bacher. 1989. Design of an expert system for environmental assessment of manufacturing processes. *Proc. Ind. Waste Conf.* 43: 425–33.

Wable, O., E. Brodard, J.P. Duguet, J. Mallevialle, and M. Roustan. 1990. An expert system for the design and sizing of ozonization reactors used in water treatment. *Eau, Ind., Nuisances.* 136: 37–8.

Wienke, D., and K. Danzer. 1986. Evaluation of pattern recognition methods by criteria based on information theory and Euclidean geometry. *Anal. Chim, Acta.* 184: 107–16.

Wood, D.M., Houck, M.H., Bell, J.M. 1990. Automated calibration and use of stream-quality simulation model. *J. Environ. Eng. (N.Y.).* 116(2): 236–49.

15

Expert System as Metaphor: The *AskARIES* Knowledgebase

Charles K. Mann

Introduction

Both to create employment opportunities and to foster more dynamic national growth, assistance to the developing countries increasingly includes programs for encouraging and assisting small and micro-enterprises. Most of these programs are implemented by intermediary institutions such as Private Voluntary Organizations (PVOs), other kinds of Non-Governmental Organizations (NGOs), development banks, cooperatives and various sorts of associations. However, the capacity of these organizations to design and implement enterprise development programs often proves to be inadequate, particularly in the case of organizations relatively new to programs of this sort. To assist intermediary organizations to improve their capacity to work more effectively with small enterprises, the U.S. Agency for International Development created the *ARIES* project (Assistance to Resource Institutions for Enterprise Support). The prime contractor for the project was Robert R. Nathan Associates and the subcontractors Control Data Corporation, Appropriate Technology International, and the Harvard Institute for International Development. The project supplied technical assistance, training and applied research, with HIID bearing principal responsibility for the applied research component.

An earlier version of this paper was presented at the American Association for the Advancement of Science symposium: "Expert Systems in Development: Advances and Applications," Boston, Massachusetts; February 13, 1988.

AskARIES was developed as a part of the *ARIES* Project (Assistance to Resource Institutions for Enterprise Support) with support from the Office of Private and Voluntary Assistance, U.S. Agency for International Development Washington, D.C. 20523. Contract Number DAN-1090-C-00-5124-00

The HIID team started by assessing the ways in which existing capacity of the client organizations was regarded as deficient. To facilitate this task, capacity was divided into four broad categories: strategic, administrative, technical, communications. Through a process of literature review and interviews with management and staff of resource institutions, information was developed on capacity shortcomings within each of these four domains. Operationally, this resulted in a sort of inventory of the problems which the institutions frequently encountered. These problems were then examined for similarities, patterns, ways in which the problems could be categorized and clustered. The result was a typology of "recurrent problems" (Mann, Grindle, and Shipton 1989).

The Creation of *AskARIES*

From the outset of the project, there was provision in the project for a computerized database to assist in organizing and analyzing the large amount of information available within the literature relevant to small enterprise development programs. The idea of "recurrent problems" proved to be a powerful concept around which to organize the database so that information could be related easily to the project's capacity development mission. Specifying problems led to inquiry as to the causes of the problems, directions in which solutions might lie, and implications for the project—particularly with respect to training. Such information was included in the database.

This definition of database structure and content conditioned somewhat the selection of software to be used. However, there were other considerations as well. The completed product was to fit easily into the existing computer environment of these organizations, be inexpensive, and be easy to learn and use. Most of them were using either IBM-PCs or compatibles and, to the extent that they had a database standard, it was Ashton-Tate's dBase. However, dBase was ruled out because at that time it could not accommodate the long text fields anticipated. With help from Harvard's Office of Information Technology (OIT), the *ARIES* team evaluated a number of products and ultimately selected Notebook II by ProTem Software. Designed as a bibliographic system with virtually unlimited field size, it was inexpensive and relatively easy to learn and use. A companion program permits embedding "keywords" in a text, e.g. (Smith 1980) and subsequently using these to trigger the automatic creation of an alphabetized and formatted bibliography of those referenced works. This feature alone was judged to make the finished database useful to anyone writing on the subject of small enterprise programs.

The Structure and Concepts Underlying *AskARIES*

Since the project's objective was to help institutions cope more effectively with their recurrent problems, these represented the central focus of the database. They were categorized and organized into a hierarchical structure, divided broadly into the four domains of capacity type. Figure 15.1 displays a record's hierarchial field structure. Note that there is provision for discussion of the problem, its causes, implications, suggestions and training ideas for strengthening capacity to deal with the problem.

In addition to capturing in the database explicit information about recurrent problems drawn from the literature, the *ARIES* team wanted to create a structure that also would facilitate exploring relationships between problems and a wide range of variables that might affect program design and success. Therefore fields were created for such things as the context within which a program is operating, various attributes of the resource institution, central features of its program, and characteristics of its clients. In addition to providing descriptive information, this structure allows formulation and testing of various hypotheses relating the recurrent problems to this range of other factors in the database. For example, one might hypothesize that the structure of interest rates and borrower fees would be substantially different according to cultural context and religion of the clients, given, for example, Islamic views on the concept of interest. The system's structure was designed to allow the testing of such hypotheses.

The Expertise and the Experts

Once the database was created, well-trained research assistants were recruited—most with field experience in developing countries, some in small enterprise development—to search through the literature for evidence of recurrent problems and information relating to their occurrence. The R/A's analysis of the literature with respect to these problems resulted in two sorts of expertise becoming embodied in the database: expertise about the literature and expertise contained in the literature. With respect to the first, the *ARIES* analysts sought out particular publications pertinent to the "recurrent problems." Out of dozens of potential sources, a relatively few were chosen as especially helpful with respect to each particular "recurrent problem". Aided by the structure of the database, the analyst made explicit the relationship between the information in the article and the real-world "recurrent problem" of the resource institutions, including insights into problem causes and implications. The second type of expertise was drawn from the literature itself, being the research findings, opinions, evaluations and conclusions of various specialists about particular programs, institutions, problems, and general principles of enterprise development and program design and management.

FIGURE 15.1 Structure of *AskARIES*

Capacity Type	Recurrent Problem	Problem Subcategory
STRATEGIC	SETTING PRIORITIES	Assessing the need Knowing the environment Considering feasibility
	BECOMING EFFICIENT	Cost effectiveness Staffing and support Centralization/decentralization
	MANAGING CHANGE	Expansion/contraction Reorientation
	CREATING INDEPENDENCE	Independence from funders Independence of clients
TECHNICAL	PROJECT DESIGN	Appropriate designs for credit and marketing projects Pricing of services Interest rates Appropriate technical assistance Sequencing of activities Participation Client selection and monitoring
	ACCOUNTING PRACTICES	Short-term budgeting Medium/long-term financing
ADMINISTRATIVE	PERSONNEL AND ORGANIZATIONAL MANAGEMENT	Hiring staff Training staff Motivating staff Coordinating staff
COMMUNICATIONS	INFORMATION MANAGEMENT	Learning from feedback Program evaluation
	INSTITUTIONAL LINKAGES	Networking with other resource institutions Linkages with governments and international donors

The Bibliographic Database

An earlier work takes note of the commonplace observation that pattern-ing computer software on a familiar metaphor facilitates its acceptance and application, witness the spreadsheet and VisiCalc (Mann 1987). Its more origi-nal contribution is the observation that the very familiarity of the metaphor tends to blind users to the vast range of ways in which the power of the comput-er program transcends the metaphor upon which it is based. (What is the pa-per-and-pencil analog to a Lotus 123 macro?) As innovative programmers expand the power of programs, connections to familiar metaphors become ever more tenuous, misleading and limiting.

This phenomenon can be seen in the development of the original *ARIES* database. The software used—Notebook II—was conceived originally as a computerized elaboration of a bibliographic card file which served as the guid-ing metaphor. Using such a database management system would allow exten-sive text storage and make possible a far more elaborate indexing system. However, as already noted, the possibility of entries in over 50 fields led to the discovery that cross-referencing and sorting could be so extensive and power-ful as to permit exploration of research hypotheses by selective database query. This discovery was made in that vast territory "beyond the metaphor".

Another discovery came when one of the client agencies proposed to use this database structure to organize and analyze their own document collection. This idea opened the prospect of the database becoming something like the "community notebook" described by Englebart and Lehtman (1988). The "recurrent problem" framework had been developed through discussions with the community of agencies working on a set of common problems they faced in serving small enterprises. Most felt some sense of "ownership" in the system. Why not all use this common framework and system to organize and extract information from their own publication collections? Once this information was on diskette, it would become a simple extension to share knowledge of their collections with other agencies. The database could become a growing information source for all of them as they as well as HIID continued to add information to it.

While this "community notebook" dimension gave the database poten-tial beyond the originating metaphor of the bibliographic card file, it was the analytical and problem solving orientation of the database structure which car-ried it far beyond the capacities of this metaphor. Whereas the typical biblio-graphic system contains only a document summary, or abstract, the structure of this one forced reviewers of documents to search out information on prede-termined topics of interest to this particular community. Figure 15.2 compares the *ARIES* system with the conventional bibliographic system. The informa-tion shown outside the shaded area of the figure normally would not be picked up in an abstract. However, having fields for key variables and for information relative to particular problems prompts the reviewer of the document to seek out actively such information: key program and environmental characteristics, program context, attributes of organizations and clients, insights on particular problems. Thus the framework itself helped to encourage within the community a common analytical approach, with a focus on applied problem solving. As so often happens when the power of the computer is applied to traditional tasks, the original metaphor is useful as a starting point, but new horizons should be sought beyond the metaphor, lest it become limiting rather than facilitating.

Having all of these attributes, the original *ARIES* system more than fulfilled original expectations. Its problem solving orientation and its potential for exploring relationships among program variables already had carried it well beyond its generational metaphor of the bibliographic database. Further expansion and development seemed to lie in adding the brute memory capacity of the CD ROM to provide the full text of key documents. In metaphorical terms, such an extension was captured superbly by Byte Magazine columnist and SciFi author Jerry Pournelle when he sketched out his vision of a CD "Library of the Month Club" in a talk celebrating The Boston Computer Society's Tenth Birthday.

FIGURE 15.2 Structure of a Typical *AskARIES Record*

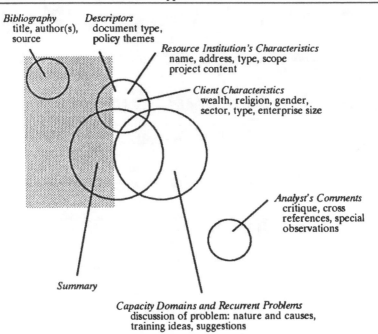

Source: Mann, Grindle, and Shipton 1989: 405.

The Paradigm Shift from
Bibliographic Database to Expert System

In writing about the nature of scientific revolutions, Thomas Kuhn describes the power of the paradigm shift to reorient thinking about scientific phenomenon. (Kuhn 1963) Looking at the *ARIES* database from the perspective of a different generational metaphor produced something of the same effect. As noted above, we, its designers, saw it as an extremely well-developed database, already pushed well beyond conventional databases. Into this complacency was dropped the question posed for the AAAS symposium: "Could this be considered an expert system?" Would that paradigm capture its essence better than "bibliographic database"? Attempting to answer these questions led to a fundamental reconceptualization of the *ARIES* database, leading to its transformation into the *AskARIES Knowledgebase.*

In spirit, the original database possessed some characteristics of an expert system. It was designed to help solve particular problems of program design and management pertaining to a relatively specific subject matter domain: small enterprise development programs. It was structured in a particular way by virtue of extensive consultation with experts and by review of the domain's literature. Users of the system use the system to help define their problem and to seek expertise pertaining to it.

While the motivation for the system had much in common with that underlying expert systems, its structure had few characteristics of an expert system. The major components of such a system are portrayed graphically in Figure 15.3. As a database, the *ARIES* product seemed actually to break some new ground. However, from the perspective of an expert system, it represented only the most rudimentary beginning. In effect, it represented a hierarchically structured knowledgebase about small enterprise programs. It had no rules, no inference engine, so user interface tailored to this particular subject matter domain.

Since the database metaphor was found to constrain an understanding of the potentials of the *ARIES* system, the descriptor "bibliographic database" was dropped and the system was rechristened the *AskARIES Knowledgebase.* This renaming was intended to signal the user that here was a product distinctly different from a conventional database. Approaching it with the database mindset would blind the user to much of its potential, particularly its problem solving orientation. Instead of thinking "I already know approximately what that is.", the new user is more apt to think "I wonder what a knowledgebase is? Using the term "database" for *AskARIES* would tend to limit comprehension of potential in the same way that the term "automatic typewriter" would have limited understanding of a word processor's potential. Greater insight comes from moving it into the context of expert systems by labelling it as a knowledgebase. "What's in a name?" A fundamentally different mindset.

FIGURE 15.3 Typical Expert System Architecture

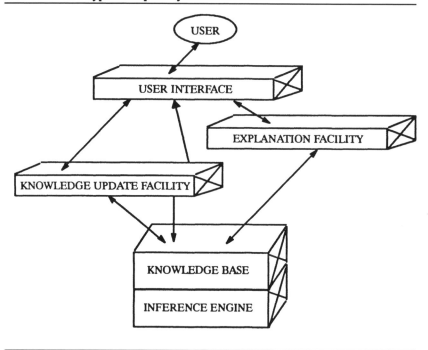

Source: Rolston 1988.

This different mindset proved to be enormously provocative, stimulating a whole new line of useful inquiry. Issues arose immediately that simply never surface with a database. Of course there were no rules embodied in the system. However, the ES perspective leads one to ask "Could there be rules? Would they be useful? For example, do experts in the field of small enterprise development have rules of thumb that guide their program design? Can some "rules" be derived from examining the information accumulated in the knowledgebase? It proved highly productive of new insights to look for rules that might be implicit in or could be derived from the summaries and analyses of published material contained in the knowledgebase. What kinds of statements can be made about various kinds of projects; about principles of project design with respect to particular contexts, resource institutions, their objectives and their clientele?

The expert system as paradigm also focused attention on the user interface. Currently, the *AskARIES* user sees only the command menu of Notebook II: *"Edit, Print, Utilities, Files, Name, System, Quit"*. There is nothing specific to the subject matter domain. One could imagine a mini-expert system embedded in an interactive user interface. This could embody some of the expertise of the analysts who helped to create *AskARIES*. There were some general social science principles that guided the selection of information for entry and its analysis with reference to the "recurrent problems". Thinking about *AskARIES* in the context of "expert system" rather than "database" encourages efforts to identify and make explicit these rules and principles. Doing so promises to bring greater consistency to the process by which the knowledgebase is created; to shorten the training time for the contributing analysts; to facilitate analysis and entry by staff of other collaborating agencies; and to help individuals use *AskARIES* more efficiently and effectively.

The new insights and effectiveness described above all flowed from shifting the perspective from which the original *ARIES* database was viewed. The expert system paradigm pushed us to seek these new insights. The database paradigm did not. In terms of the generational metaphors of the two paradigms, the database has its roots in the vision of the filing system; the file drawer, the card index file: passive repositories of information. The generational metaphor of the expert system is the active, problem-solving expert. Shifting to this latter context encourages a completely different mindset. One seeks to identify and to make explicit for users of the system the rules of thumb, the guidelines, the general principles which guide the real experts as they approach a particular set of problems.

As to *AskARIES*, this expertise exists at two levels. At one level, it relates to the rules that guided the analyst expert in designing the system; rules that can help a user to use the knowledgebase itself more productively as an information resource. At a more fundamental level, the expertise relates to the rules and principles which are implicit in the information contained in the entries themselves; rules which are used by experts in small enterprise development programs, knowledge of which can help non-experts improve their program design and management performance. Are there such rules of thumb and general principles which can be made explicit? The attempt to discover them will produce a better understanding of the subject of small enterprise development.

Yet another feature of expert systems that prompts useful extensions of *AskARIES* is the common ES feature of being able to explain the "reasoning" by which it reaches particular conclusions. Incorporated in *AskARIES* this feature would allow the user to compare the rationale embodied in the system with his or her own ideas and expertise. Such a capability also would enhance its value as a learning tool for newcomers to the field. The rules-of-thumb and principles used by experts in the field would be made more explicit than they are in a bibliographic database.

Another way in which the reorientation toward ES opened new potentials was in liberating the system from its original exclusive reliance on publications or existing AV materials as the source of the knowledge. This reliance was intrinsic to the concept of a "bibliographic database". The sources of knowledge in an ES knowledgebase typically are not publications as much as they are interviews with the experts themselves. This reorientation caused a fundamental reconceptualization of sources of knowledge to be tapped in future extensions of *AskARIES*. This ES vision, for example, suggests interviewing the experts about their approach to particular problems, and provides a portfolio of "knowledge engineering" techniques to extract and codify expertise. Video clips also could have a place in this knowledgebase, "made to order" to illustrate particular techniques, situations, or lessons learned. With computers already able to digitize both audio and video, such entries could be included directly, instead of by reference. Such extensions are more apparent through the ES lens than through a DB lens.

Conclusion

In conclusion, the paradigm shift from bibliographic database to the knowledgebase component of a potential expert system was facilitated by the fact that the original database had been organized around a series of "recurrent problems". Therefore, it already embodied the problem-solving orientation of an expert system, if not its form. Moreover, the analysts working on the original database had applied substantial expert knowledge as they scoured the literature of small enterprise development for information about these recurrent problems and avenues leading to solutions. They were seeking to dig out information relevant to specific problems; not trying to summarize publications (although summaries are included in *AskARIES*).

Once it occurred, the paradigm shift reinforced the problem solving orientation of the database and diminished tendencies for the system to become an ever more richly annotated bibliography. As noted above, the new paradigm suggested moving beyond published information and trying to capture more expertise directly from interviews with leading practitioner experts; the so-called knowledge engineering of expert system development. While such a process of systematic interviews remains for the future, numerous smaller changes have helped to sharpen the problem solving orientation of the system; making it more explicitly a tool for "seeking solutions" to problems of small enterprise development.

Whether the expertise concerning such a diverse subject as enterprise development could be captured even by a highly sophisticated expert system remains a subject for future investigation. However, many tangible improvements to *AskARIES* have resulted from reconceptualizing it with an expert system perspective. Even if no further extensions are undertaken, the reorientation already has enhanced its usefulness both as a reference tool for problem-solving managers and as a training tool for newcomers to the field of small enterprise development.

References

Engelbart, D. and Lehtman. December, 1988. Working together. *BYTE*, 245–252.

Kuhn, T.W. 1963. *The Structure of Scientific Revolutions.* International Encyclopedia of Unified Science, Vol. 11, No. 2. Chicago: The University of Chicago Press.

Mann, C.K. 1987. Beyond the metaphor: microcomputers in public policy and human capital development. In *Microcomputers in Public Policy; Applications for Developing Countries.* ed. S.R. Ruth and C.K. Mann. Boulder: Westview Press and the American Association for the Advancement of Science.

____, M. Grindle, and P.M. Shipton, eds. 1989. *Seeking Solutions: Framework and Cases for Small Enterprise Development Programs.* West Hartford, CT: Kumarian Press.

Pournelle, J. *An evening with The Boston Computer Society.* The Powersharing Series. No. 143, Riverside, CT: Powersharing Press. Audiotape.

Rolston, D.W. 1988. *Principles of Artificial Intelligence and Expert Systems Development.* New York: McGraw-Hill.

Waterman, D.A. 1986. *A Guide to Expert Systems.* Reading, MA: Addison-Wesley Publishing Company.

16

Future Promise in Light of Current Practice

Victor S. Doherty,
Charles K. Mann, and John J. Sviokla

The Structure and Application of
Agricultural and Medical Knowledge

Expert systems have been applied so far to relatively few of the problems of developing countries. The cases described in this book, however, suggest some of the possibilities, and some of the benefits to be expected. The most promising knowledge domains include those that form the main subjects of the previous chapters: agriculture and health care. Both fields involve knowledge that is reasonably well codified, is in the hands of relatively few experts, and is intended for application on a massively large scale. Expert systems are particularly well-suited to diagnostic applications, and diagnosing problems is a key function of experts in both agriculture and health care.

As between these two fields, the tally of systems presented in this book is heavily weighted to agriculture. Within the extensive system of U.S. agricultural landgrant universities, for example, several institutions appear to be seeking as a matter of policy to take leading positions in the application of new information processing technology. The agricultural engineering departments in many landgrant institutions have been particularly active, and there has been strong interest on the part of many extension departments as well. In addition there has been considerable work in other countries on the agricultural application of knowledge-based technology.

Such widespread and far-advanced activity is not limited to agriculture, however. Some of the earliest and most successful expert systems were in medicine. These early medical systems continue to be widely influential as logical models and as software resources for use in and out of medical fields. Some medical work has been discussed in the preceding chapters, and medicine continues to be an area of extensive activity and progress for work in knowledge-based systems.

There are important differences, however, in the traditional ways in which agricultural vs. medical knowledge has been handled. These differences are still of a widespread, structural nature. A review of some apparent reasons for this structure suggests that it may be reasonable to expect the operational, field use of agricultural expert systems to precede operational health care applications in the developing countries. This appears to be the case even though both subjects are of critical importance, and even though field work in applications relating to both areas is of high priority.

One source of structural differences between agricultural and health care knowledge appears to be that in agriculture there is a policy emphasis on getting information to large numbers of individual farmers, and on convincing them of the usefulness of this information. This emphasis contrasts with that in health care, where instead of relying on individuals' ability to treat themselves the focus is on getting information to practitioners, who then treat large numbers of individuals.

Emphasis on disseminating agricultural knowledge to individuals has grown rapidly along with the growth of science over the last two hundred years; it has accelerated as a result of recent breakthroughs, including the development of hybrid maize during the first part of this century. As agricultural historians are well aware, however, technology dissemination and adoption have been matters of considerable political moment from even earlier times, and evidence of this importance has been found throughout the world. Examples could be multiplied from many of the oldest agricultural civilizations—including Egypt, Rome, and the early states of the Gangetic plain—where historical documents show that agricultural policies and production schemes have been matters of concern and planning to kings and governments for a long time.

For many thousands of years almost every man or woman in a post-neolithic society has been at least a potential farmer, and most have been farmers in fact. Literally since the days of the first shamans, however, health care beyond the limited extent of home remedies and traditional ameliorative measures has been the domain of specialists. The steady multiplication of agricultural knowledge and alternatives has meant that farmers need either to learn more or to specialize, but it has not changed the fact that individual farmers make their own decisions however complex or fraught these decisions may be. By contrast, individual persons who need health care are likely to seek out a specialist if possible, for any problem of importance that cannot be dealt with on the level of home remedies and home traditions.

The reasons for this long-standing structural difference in how knowledge is held and applied seem to run as follows. A farmer needs a thorough understanding of the agricultural situation, but the set of conditions and factors dealt with repeats itself year after year, affecting familiar crops and soils, in familiar ways that are at least roughly predictable. Variation in conditions and in crop responses is important for farmers, but most can see in their own or their neighbors' fields the major part of the variations that are likely to be important to them. Medical problems are much less likely to be predictable, and medical problems are much more unevenly distributed. Even the traditional practitioner in a preindustrial situation needs wider experience than the farmer, and even in a technologically simple society there are important alternatives available for the treatment of illness, covering a wide range including herbal, psychological, and dietary interventions. The farmer knows what the round of the seasons will be, but it is not at all easy to predict illness, nor is it possible without apprenticeship and study to see enough health problems in enough detail to support useful understanding.

Such considerations have produced the situation that we still have today, in which knowledge about health is much more likely to be applied by a specialist than is knowledge about agriculture. The rapidly expanding range of agricultural possibilities in today's world means that farmers must seek outside help more often, and it is a major point of this book that expert systems can be of vital help in expanding agricultural horizons beyond the boundaries and conditions of a single village. The requirements of dealing with health care, however, combine to force a situation in which, at the level of the town and the village and the home, the doctors and health workers who apply medical knowledge to care for individual persons are and will be fewer than the farmers who apply agricultural knowledge to individual fields.

Often those medical expert systems that are fully operational in the developed countries are focused on particular diseases or disease complexes; often the software systems are relatively large, and commonly they are applied in hospital situations. As we have seen in the preceding chapters, expert systems for even simple modern health interventions by rural health workers and similar practitioners need more work and testing than has been done so far.

Field health workers who are not doctors will come to be more and more important, and will be aided very greatly by expert systems. As between health care and agriculture, however, operational applications of expert systems support for health care in developing countries may have to be approached more slowly. It appears to be in agriculture that much of the first pioneering work will be done, in learning how expert systems can be used in a widespread, fully operational way to deliver information and to turn it widely into knowledge held and applied by large numbers of people.

The Promise of Agricultural Expert Systems

A farmer is, in effect, the proprietor of a factory for food, with tremendous amounts of local variation in product and process. Because of this situation, farmers need up-to-date, effective information about the alternatives they have and about the reasons for choosing among them.

Much of the knowledge that applies to planning and control of the agricultural process is tacit—it is locked in the heads of experienced farmers. Other knowledge is implicitly shared—it is embedded in the social context, in contacts within a network of farmers, customers, suppliers, and scientists. Knowledge-based systems can help individuals in these networks to extend, expand, and improve their knowledge and contacts. For both developing and developed countries, this can bring

- More productivity and employment in agriculture, through wider and more diverse application of new scientific results.
- Wider scope for individual, managerial initiative by all farmers: reinforcing local abilities, to solve local problems.
- Environmental improvement: making available the knowledge, and the connections with new markets, that will provide incentives to invest in the natural environment rather than mining it.

Figures 16.1 and 16.2 diagram these points. Increasing and assuring the world's food supply, and doing so in a way that allows farmers to get ahead of the poverty curve and stay there, means building a developed agriculture: one in which the countryside receives from the city as well as supplying it, and in which unskilled or semiskilled labor becomes skilled labor that is expended on complex production and marketing systems.

Critical Knowledge Areas for Agricultural Information Delivery

In the last few decades we have witnessed a revolution in the creation of new agricultural knowledge, and in harnessing new means to apply it. Each aspect of this changed situation increases the importance of information, information which is no longer available simply by observing one's own and one's neighbor's fields. Developments such as the growth of biotechnology mean that the degrees of freedom that farmers have to work with are increasing and are likely to be multiplied severalfold within the relatively near future. Major areas in which extensive new knowledge is available, and in which information is becoming a critical factor of production in its own right, include:

- New, fertilizer- and management-responsive crop varieties that have become of worldwide importance.

FIGURE 16.1 Current Situation

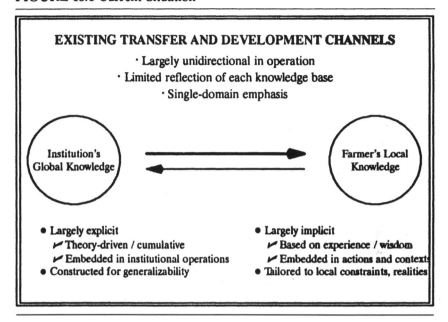

EXISTING TRANSFER AND DEVELOPMENT CHANNELS

· Largely unidirectional in operation
· Limited reflection of each knowledge base
· Single-domain emphasis

Institution's Global Knowledge

Farmer's Local Knowledge

● Largely explicit
  ✔ Theory-driven / cumulative
  ✔ Embedded in institutional operations
● Constructed for generalizability

● Largely implicit
  ✔ Based on experience / wisdom
  ✔ Embedded in actions and contexts
● Tailored to local constraints, realities

FIGURE 16.2 Projected Situation

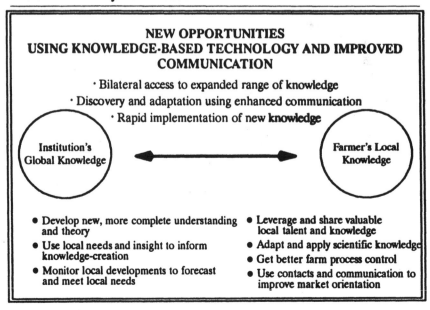

NEW OPPORTUNITIES
USING KNOWLEDGE-BASED TECHNOLOGY AND IMPROVED COMMUNICATION

· Bilateral access to expanded range of knowledge
· Discovery and adaptation using enhanced communication
· Rapid implementation of new knowledge

Institution's Global Knowledge

Farmer's Local Knowledge

● Develop new, more complete understanding and theory
● Use local needs and insight to inform knowledge-creation
● Monitor local developments to forecast and meet local needs

● Leverage and share valuable local talent and knowledge
● Adapt and apply scientific knowledge
● Get better farm process control
● Use contacts and communication to improve market orientation

- Practices such as integrated pest management to reduce both the costs and side effects of production inputs.
- Wider and more complex markets, requiring farmers to have the information and knowledge to respond quickly.
- Increasing scope for choice among available, locally-adapted new varieties.
- Growing importance of multiple cropping and other management-based strategies for increased production.

Traditionally, the transfer and adoption of new agricultural technology has traced a gradual, s-shaped curve when the percentage of adopters was plotted against time. Once farmers and extension agents accept the concept of using expert systems for guidance in agricultural decision making, choice of technique, and so forth, there will exist a powerful vehicle for accelerating the pace of future changes. Knowledge about new technologies will be able to move more quickly to those farmers for whom they are appropriate, and the s-curve of adoption should become progressively steeper, even approaching the form of a step function in the best cases.

As new knowledge is developed, diskettes with improved expertise can be supplied as updates to systems, much as new versions of popular software packages replace older ones. The concept of using expert systems would be familiar already to both farmers and extension agents.

Developments in Hardware and in Policy

Continued miniaturization and the continued introduction of other new technology will complement and expand the potential of expert systems. Developments including the smart card, for example, promise rugged, hand-held computers for field use; digital and cellular communications mean that users of such machines in remote sites can use them to interact with a central data base (Lissandrello 1990).

Applications in Africa of microcomputers linked through cellular communications suggest how new technology can overcome the expense and maintenance problem that current, cablebased technology presents under conditions of extreme climatic stress and long geographic distances (Madsen 1990). In Southeast Asia, countries including Malaysia (Fong 1989) plan heavy investment in telecommunications infrastructure during the coming decade. These plans embrace both cable and cellular installations, and are being made in the context of rapid, extensive national and international communications developments that will greatly expand regional links within Southeast Asia and in the Pacific area as a whole (Hukill and Jussawalla 1989; Barber 1989).

Networked communication is an area of particularly active planning and development in Asia. In Malaysia, where agricultural employment and the standard of living of agricultural families are of high political importance (Hitam 1984, e.g.), work has begun at the national agricultural university (Universiti Pertanian Malaysia) to develop a country-wide agricultural information network that will provide information about particular crops to farmers, merchants, scientists, and goverment administrators. Similar communication networks for agriculture are in place or under planning and development in many areas in the United States. Countries such as France and Japan, where national policy mandates the development and country-wide dissemination, down to the individual household, of computer-based telecommunications and information processing can be expected to participate actively in this area of work. These rapidly expanding national data links open the prospect of expert systems at the user's site tapping into large central databases for input: for example, to bring weather forecast data into a crop management expert system.

Producing, Organizing, and Distributing the Information

The major generators of scientifically evaluated and organized information about crops and cropping patterns are public institutions: universities, government research stations and institutes, and international research institutes. Most of the expert systems described in this book come from such institutions. In addition, private organizations including seed companies, fertilizer companies, agrichemical firms, and biotechnology firms conduct research and extension work that is focused on particular products of the firm, and on their use.

In addition to classical extension work by public and private institutions, new models for dissemination are needed. Book, magazine and software publishing can provide such models. Publishers are already heavily involved in turning new information technology to creative use for teaching and learning. For example, the *AskARIES Knowledgebase* (Chapter 15) was published by a traditional book publisher that recently redefined its mission not as the publication and marketing of books, but as the publication and marketing of information, without regard to the publishing medium. Increasingly, magnetic and optical (CD) media may replace paper in publishing large bodies of information. In the case of information technology and services for farmers, the alliances that publishing companies are building with software and hardware companies, and with research and teaching institutions, need to be strengthened and developed.

The division of labor that has arisen within book publishing serves as a guide to ways in which the agricultural information systems field can develop, specifically to provide a vehicle for disseminating expert systems as they become available. Textbooks as well as trade books on technical subjects currently are written largely by university faculty members, on contract or on some other arrangement. The university gets the benefit, for its teaching function, of faculty members' organizing and developing their insights and presentation during the writing process. The faculty member gets a lump sum payment, or royalties, and recognition that is useful in salary and job negotiations. The publishing company makes its profits by organizing and supporting the production and distribution of the books. All three groups are able to benefit: because they are involved in the production and distribution of knowledge that is widely sought, but is impossible to disseminate broadly using only the the author's direct lectures or teacher-student interaction.

Building new information systems for farmers, then, involves two main steps:

- Developing the electronic equivalent of book and magazine publishing for information about crops, technology, and markets.

- Developing, distributing, and operating hardware systems to support access to this information, by all farmers.

These considerations lead directly to the identification of a series of potential actors, and to the identification of their interests. From the list of these actors that follows below, it is obvious that formal, institutional structure will vary from country to country. Communications companies are a prime example, where in different countries one meets different structures ranging from government monopoly through closely regulated utility to startup, relatively unregulated enterprise. Also, particular actors and interests noted below can be combined; university presses are an obvious point at which publishing and research interests converge, and some of the landgrant universities in the United States are already moving to develop and market knowledge-based systems and similar software. Nevertheless, the list of groups likely to be important in bringing expert systems to the operational levels and widespread importance envisioned in this book appears to include:

- Farmers' organizations: groups with an interest in seeing expert systems applications come about, and in shaping their structure. These groups include general membership farmers' organization, as well as marketing, processing, and production cooperatives.

- Publishers: companies whose major product is information, and whose interest lies in having as many people as possible use their products.

- Communications utilities: whatever a country's institutional structure in this area, communications utilities will be key participants in bringing about the wide uses of expert systems that are argued for here.

- Computer hardware designers and manufacturers: providers of the machines that drive and make possible an information delivery effort.
- Software designers, builders, and manufacturers: with hardware manufacturers, they provide the information medium.
- Universities and research institutions: providers of scientific knowledge, systematically developed and presented. Like publishers, they benefit when more people use their product; in addition, they have a public mandate to create new knowledge.
- Seed, biotechnology, fertilizer, and agrichemical companies: groups with a primary interest in marketing crop production materials, and in having farmers know how to use these effectively.
- Governments: representatives of the public interest.

Research on specific expert systems applications and on the results of their use are matters of high priority. In different countries, there will be different institutional roles for universities, national research institutes, and other groups. On a world scale, the research institutes of the Consultative Group for International Agricultural Research (CGIAR) are particularly important: these institutes have international funding and responsibility for research on particular crops and crop production environments, which gives them a worldwide leadership role and means that there is worldwide attention for the results of work that they carry out alone or in collaboration with other organizations.

Several of these institutes, as well as a number of university-based groups, have made important strides in modelling the growth and environmental reactions of basic crops such as rice, maize, and sorghum; these models provide important starting points for the application of expert systems and related software technologies to the improved delivery of information to farmers.

What Questions Should Be Asked

The highest priority is to move to the field as soon as possible. There are several questions that can only be answered on the basis of field data, and that will determine the success of expert systems' application to agriculture. One of the most important of these questions is what kind of focus will be most useful to farmers: systems dealing with particular crops or pests; or systems with a more general agronomic focus, for example, designed to elucidate the range of possible uses for a particular type of soil and climatic situation.

Questions of software strategy are also of great importance: which will be of the most use to farmers, fully-fledged expert systems oriented to solving specific problems and performing specific planning tasks; or more flexible, open-ended data and inquiry structures that come closer to hypertext? From a macro-level point of view, we need to know what type of system and what strategy for system deployment will have the most effect overall on farmers' levels of useful knowledge.

Finally, and very importantly, we need to know to what degree information dissemination should be combined with communication—among farmers, between farmers and customers, and between farmers and scientists. Well-tested and well-developed research methods—in social network analysis, psychology, learning, farming systems, and village studies—exist to help in framing the questions and in gathering and analyzing the data.

Conclusion

ES in Technology Transfer

Although expert systems are rapidly assuming key roles in both production and service industries in the developed countries, it is still too early to judge which of the two statments that open this book will prove to be more correct in the field of general economic development: that expert systems will have few practical applications, or that they will play a major role in the dissemination and application of useful knowledge leading to economic growth and higher standards of living. Nevertheless, the pioneering efforts documented in this volume demonstrate some concrete ways in which ES can be used within developing countries. Moreover, the existence of flexible, inexpensive, user-friendly ES shells means that applications can be created widely, within the developing countries themselves. Thus, not only does ES have the potential to expedite technology transfer from North to South, but it also promises to multiply the impact of the developing countries' own experts. Relatively simple, home-grown systems may prove to be an excellent knowledge resource to enhance the effectiveness of paraprofessionals in a wide array of fields, as in the case of the system for village health workers described in Chapter 12.

ES in Capacity-Building

In closing, we note the potential role that expert systems could play in capacity-building programs. Potential for capacity-building effects exists at the point of knowledge generation as well as at the point of knowledge use. At the first of these points, that of knowledge generation, one benefit of constructing an expert system is the discipline it imposes on the expert to clarify and systematize the knowledge that is to be incorporated into the system. This process should help to shape the knowledge into a form more easily communicated to others; clarifying the expertise itself may be an important by-product of the process of building an expert system.

At the point of knowledge use, consider the example of an extension worker or a farmer using an ES regularly to help diagnose plant diseases.

Using the system over a wide range of conditions and having it display its "reasoning" in each case, the user should begin to develop the same diagnostic patterns of thought characterizing the original expert, whose knowledge and strategies for analysis are embodied in the system.

A well-constructed ES should function as a sort of tutor for the regular user of the system, a function that should be present and important both in health care and in agriculture. Thus, expert systems are not only vehicles to apply expert knowledge to particular problems, but are potentially powerful learning resources to help ES users to develop their own expertise. This capacity-building potential may be one of the most important aspects of the spread of expert systems within the developing countries.

References

Barber, R. 1989. Pacific telecommunications: The role of regional organizations. *Columbia Journal of World Business*. 24(1): 101–103.

Fong, C.O. 1989. The Malaysian telecommunications services industry: Development perspective and prospects. *Columbia Journal of World Business*. 24(1): 83–100.

Hitam, Dato' M. 1984. Opening address. In *Improving Extension Strategies for Rural Development*, ed. S.M. Yassin, I. Beavers, and I. Mamat. Kuala Lumpur: Universiti Pertanian Malaysia Press.

Hukill, M.A., and M. Jussawalla. 1989. Telecommunications policies and markets in the ASEAN countries. *Columbia Journal of World Business*. 24(1): 43–57.

Lissandrello, G.J. 1990. Smart cards and handheld terminals as applied to field communications. *Telematics and Informatics*. 7(2): 101–108.

Madsen, W. May 1, 1990. African nations emphasizing security. *Datamation*. 104–2, 104–3, 104–8.

About the Contributors

Le Istiglal Amien is the Director of the Agroclimate Division of the Center for Soil and Agroclimate Research, Bogor, Indonesia. Dr. Istiglal has participated in expert systems workshops as well as written several systems in current use.

Howard Beck is an Assistant Professor in Agricultural Engineering at the University of Florida. He has degrees in Electrical Engineering and Philosophy from the University of Illinois and a Ph.D. in Computer and Information Sciences from the University of Florida. He is currently working on both practical and theoretical aspects of object oriented database structures.

Andreja Cizerle is a chemical information specialist, involved in software development and applications.

Pavel Čok, a chemist, is a postgraduate student in chemical informatics and contributed to the research for Vrtačnik's chapter.

Carol Colfer, an anthropologist, was first farming systems researcher then team leader on the TropSoils-Indonesia Project in Siting, W. Sumatra.

John A. Daly is Science Program Director, Office of Science Advisor, AID. He managed developing-country research grants programs in the 1980's. In the 70's he worked as a health planner in WHO (Columbia), the DHEW Office of International Health, and a White House international health study. In the 1960's he worked in computer research and design, and was a Peace Corps Volunteer in Chile. He has an M.S.E.E. and Ph.D. in Administration.

Victor Doherty is an anthropologist with a major interest in new information technology and its use in agriculture. He is concerned with technology and the organization of work, with innovation, and with patterns of formal and informal cooperation and agreement. His experience includes work at the International Crops Research Institute for the Semi-Arid (ICRISAT), on village-level studies of agricultural development; and at the Harvard Business School, with a project to examine ways in which new information technology is being used by managerial groups.

Dasnica Dolničar, a mathematician, is a postgraduate student in chemical informatics and participated in the research for Vrtačnik's chapter.

David Mendez Emilien is currently completing his Ph.D. degree requirements in Management Science at Michigan State University. He holds

Masters degrees in Resource Economics, Operations Research/Systems Science, and Applied Statistics, all from Michigan State University. He was previously a systems analyst for the Minister of Agriculture in the Dominican Republic and was responsible for the maintenance of the database for the Geographic Information System.

P.A. Everett is a graduate student in the Computer and Information Sciences Department at the University of Florida. She has a B.S. in Agronomy and currently is working for the South Florida Water Management District in West Palm Beach, Florida.

Thomas W. Fermanian is Associate Professor of Turfgrass Science in the Department of Horticulture at the University of Illinois. His current research interests are the development of knowledge engineering tools for personal computers, qualitative data analysis for agricultural domains, crop modelling and simulation, and development of cost effective turfgrass management systems.

Sasa A. Glažar is Assistant Professor of Chemistry at the Teacher Training Faculty, University of Ljubjana, Yugoslavia. He is also a member of the research team at the Faculty of Science and Technology, working on the conceptual design of information systems and their applications in solving environmental problems.

Severin V. Grabski is an associate professor at Michigan State University where he currently teaches undergraduate and graduate information systems courses. He holds a Ph.D. in Business Administration from Arizona State University. His work has appeared in such journal as *MIS Quarterly* and *The Journal of Information Systems*, and he has made numerous presentations at regional and national professional meetings.

James Hanson currently works as a research associate at the University of Hawaii, Department of Agronomy and Soils Science programming a multiple-cropping systems simulation model. He contributed the economic component of the ACID4 while doing thesis research on aluminum-organic residue interactions in acid soils.

M. Hoberstorfer holds an M.D. and is currently enrolled in his residency to become a general practitioner.

Stephen Itoga is an Associate Professor in Information and Computer Sciences, University of Hawaii at Manoa. Research interests are expert systems, databases, and logic programming.

James W. Jones is a Professor in the Agricultural Engineering Department at the University of Florida, Gainesville, Florida. He has twenty years experience in mathematical modelling and computer simulation of crops and soils and the integration of agricultural models with expert systems for tactical decision making in crop production systems. He has helped design and develop a decision support system for agricultural technology transfer (DSSAT). He teaches a graduate course on Agricultural and Biological System Simulation and has organized and lectured in short courses and workshops in various countries. He has B.Sc., M.Sc. and Ph.D. degrees in Agricultural Engineering from Texas Tech, Mississippi State, and North Carolina State Universities, respectively.

Pierce Jones is an Associate Professor in the Agricultural Engineering Department at the University of Florida. He has degrees in Agricultural Engineering, Astronomy, and a Ph.D. (1981) in Mechanical Engineering. He is currently working with storage and retrieval of large heterogenous databases on CD-ROM.

Phoebe Kilham is a former graduate student in Geographical Information Systems and editor of the ACID3B, ACID4, and ADSS user's manuals.

Eckhard F. Kleinau is Assistant Professor of Health Management and Policy at the University of New Hampshire and S.D. candidate in Health Policy and Management and in Epidemiology at the Harvard School of Public Health. For ten years Dr. Kleinau has gathered management experience as a public health expert in curative and primary care programs in Africa and Asia. He has holds an M.S. in Health Policy and Management and an M.S. in Epidemiology from Harvard University. He received his M.D. from Eberhard Karls University in Tuebingen, Germany.

Zhi-Cheng Li is a Research Associate for the TropSoils Project, USAID with a background is in computer science. Research interests include AI applications in Agriculture and natural resource management.

Charles K. Mann is Research Associate and Lecturer in Economics at the Harvard Institute for International Development. He works mainly with government agencies in the developing countries on issues of food and agricultural policy and macro economics. Formerly with the Rockefeller Foundation, his continuing interest in computer use began when he led an RF project in Tunisia in 1982 that introduced microcomputers into agricultural planning there. This is the second book that he has edited with Steve Ruth on the use of computers in developing countries. He also was project leader for *AskARIES*, a computer-based information system published by Kumarian Press. He holds a Ph.D. in economics from Harvard University.

Ryszard S. Michalski is PRC Chaired Professor of Computer Science and Director of the Center for Artificial Intelligence at George Mason University. He is a pioneer in the field of machine learning. His AQ and NEWGEM algorithms are the heart of *AGASSISTANT*. Over twenty-five scholars from 14 countries have visited the center to collaborate with him.

P.K.R. Nair is a Professor of Agroforestry at the University of Florida. He has published internationally on many aspects of agroforestry and has twenty-five years of experience in more that forty developing countries.

Radojka Olbina, environmental engineer, is a researcher involved in the design of information support for waste minimization and waste management, including computer simulations and modelling.

Bernard Pfahringer, a research assistant at the Dept. for Medical Cybernetics and Artificial Intelligence at the University of Vienna, and a research fellow of the Austrian Research Institute for Artificial Intelligence, has been engaged in several AI projects since 1985. His research interests are in knowledge representation, machine learning and logic programming. Pfahringer received his Dipl.-Ing. in computer science from the Technical University of Vienna in 1985.

G. Porenta holds a Ph.D. in Mathematics and Computer Science and also finished his medical education with an interest in Medical Cybernetics. He conducts research projects, including an application of Computer Science and Artificial Intelligence in Medicine. After three months medical assignment in Ethiopia, he initiated the research published in this book. He is currently conducting a fellowship in cardiology and is head of the working group "Computer in Cardiology" at the Department of Cardiology, University Hospital, Vienna.

Ahmed A. Rafea is the Director of the Expert Systems for Improved Crop Management Project which is currently being executed in Egypt. Dr. Rafea has worked at Cairo University, San Diego State University, National University-San Diego, and American University in Cairo as a Professor of Computer Science.

Stephen R. Ruth is Professor of Decision Sciences and MIS at George Mason University. He is the founder of the university's Expert Systems Development Center and also director of the International Center for Applied Studies in Management Information Systems. He consults with governmental and business organizations in Europe, Africa and South America in addition to US clients, and is currently directing two projects in Eastern Europe sponsored by the Mellon Foundation. With colleague John Redmon he is also directing a new expert systems project for the United Nations aimed at improving the quality and efficiency of large scale project design and planning.

Kristopher Sprague is a software consultant and knowledge engineer who works in Centerville, Virginia.

John J. Sviokla is Assistant Professor at the Harvard Business School, where his work focuses on improving managers' ability to use information technology to increase the effective use of organizational expertise. In this context he has investigated issues affecting the use, planning, implementation, and evaluation of artificial intelligence and other information technologies.

R. Trappl is Professor and Head of the Department of Medical Cybernetics and Artificial Intelligence, University of Vienna, Austria. He is also Director of the Austrian Research Institute for Artificial Intelligence.

Metka Vrtačnik is Assistant Professor of Chemistry at the Faculty of Science and Technology, University of Ljubjana, Yugoslavia. Her research fields are chemical information systems and building knowledge bases for expert systems, including their applications in development.

Merrill E. Warkentin is an Assistant Professor of MIS in the Department of Decision Sciences and Management Information Systems at George Mason University. Dr. Warkentin teaches and conducts research in the areas of information system security management, knowledge engineering/systems analysis, groupware, and applications of decision support system (DSS) and expert systems. Along with Stephen R. Ruth, Professor Warkentin directs the GMU Center for Business Expert Systems Research, where his most recent work focuses on developing effective knowledge acquisition methodologies.

Russell Yost is a Professor of Soil Science, University of Hawaii researching and teaching in Tropical Agriculture. He is currently developing expert systems on fertilizer phosphorus management, alley farming system design, and the integration of expert systems and GIS technology.

Index

Printed and bound by CPI Group (UK) Ltd, Croydon, CR0 4YY

23/10/2024

01778240-0009